ENDANGERED ANIMALS DICTIONARY

An A to Z of threatened species

Author Clint Twist
Managing Editor Ruth Hooper
Art Director Ali Scrivens
Art Editor Julia Harris
Production Clive Sparling
Consultant Zoologist Valerie Davies
Illustrators Robin Bouttell (Wildlife Art Ltd), Stuart Carter (Wildlife Art Ltd), Sandra Doyle (Wildlife Art Ltd), Peter Scott (Wildlife Art Ltd), Myke Taylor (Wildlife Art Ltd)

Created and produced by
Andromeda Children's Books
An imprint of Pinwheel Ltd
Winchester House
259-269 Old Marylebone Road
London
NW1 5XJ
UK
www.pinwheel.co.uk

ISBN 1-861991-01-0

10 9 8 7 6 5 4 3 2 1

Printed in China

Information Icons

Throughout this dictionary you will see two icons, a globe and a series of exclamation marks, next to each entry. These will give you more information about each creature listed.

Globes These will show you where the remaining populations of each creature can be found in the world. Small red dots clearly show the locations.

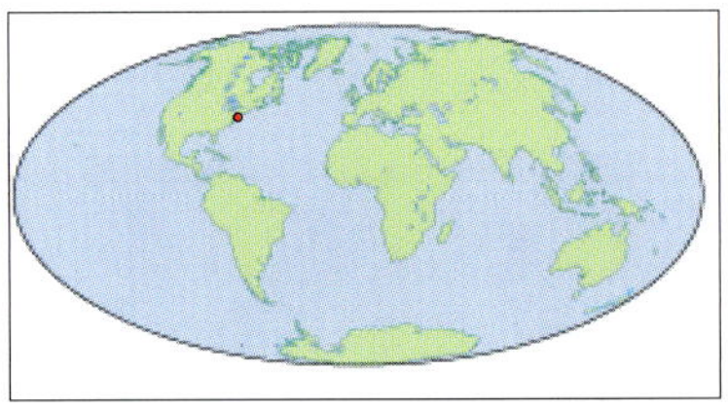

Threat of extinction The exclamation marks will tell you how serious the threat of extinction is for each creature, on a 5 level scale.

!!!!! **Extinct** - Thought to have died out, or surviving only in captivity, with no population in the wild.

!!!! **Critically endangered** - At **high** risk of extinction, with a very low or quickly declining population.

!!! **Endangered** - At **considerable** risk of extinction, with a low or declining population or a limited range.

!! **Vulnerable** - At risk of extinction, but not on an immediate timescale, mainly due to threatened habitats.

! **Lower risk** - Monitored but not judged to be threatened. Species with too little data available are also classified at lower risk.

Size Comparison Pictures

Throughout this dictionary you will see a symbol, either a hand or a man, next to a red icon of each creature listed. The hand or man will help you imagine the size of each creature in real life.

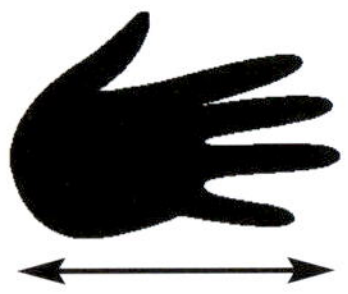

18 cm

The first symbol is a human adult's hand, which measures about 18 cm (7 inches) from the wrist to the tip of the longest finger. Some creatures are smaller than this, so the size comparison will help you imagine their size.

1.8 m

The second symbol is an adult human. With his arms outstretched, his armspan measures about 1.8 m (6 feet). The symbol will help you imagine the size of some of the really big creatures.

ENDANGERED ANIMALS DICTIONARY

An A to Z of threatened species

Andromeda Children's Books
An imprint of Pinwheel Ltd.

Trials of Life

Variety of life

Queen Alexandra's birdwing butterfly

Planet Earth is home to an incredible variety of animal life. Each different animal is a separate species, and these are arranged in groups according to their basic body design. There are about 4,600 mammal species, one of which is human beings. There are nearly 9,700 bird species, about 8,000 reptiles, some 4,700 amphibians and 25,000 fishes. There are literally countless other species as well. Scientists estimate that there are as many as 10 million insect species, of which only about 1.5 million have been discovered. Some animal species are found all over the world, while some are found only in one very small area of our planet.

Extinction

Extinction happens when a species dies out. This is a tragic event – a unique life-design is gone forever. But extinction can be a natural event, part of the long-term rhythm of life on Earth. When many species become extinct in a short space of time, scientists call it a mass extinction. Most of the mass extinctions have been caused by major catastrophes. Some scientists believe an asteroid hit Earth 65 million years ago, killing all the land-based dinosaurs. However, there may be other reasons for a mass extinction, one of which might be human interference.

Wisent

Californian condor

Human contribution

Today, the rate of extinction is very high, especially among mammals and birds. Some scientists believe this is caused by human activity. People have always hunted animals for food, or for their fur, feathers, horns and tusks. Some have been hunted to extinction, like the dodo, a flightless bird that died out in the 17th century. Hunting still causes problems, but more damage is done by human activities that indirectly threaten animals.

Galapagos penguin

Habitat destruction

Wild animals need wild habitats, and the biggest threat faced by wild animals today is habitat destruction. Humans cut down forests for timber and clear land for farmers. Grasslands are ploughed to grow crops. Hotels and marinas are built along coastlines, rivers are dammed and valleys flooded. Roads and pipelines reach even the remotest areas. All of this disturbs the delicate balance of nature. Pollution also gets into streams and rivers, and some of it falls back to Earth as acid rain. And wherever people go, their domesticated animals go with them. Sometimes these domestic species can have a harmful effect on wildlife.

Kinkajou

What Can We Do?

Publicity and education

Northern right whale

Despite the number of extinctions, humans can be a force for good where animals are concerned. In the 20th century, groups of caring people really did 'Save the Whale' and whale numbers are now on the increase. The most important weapons in the fight to save endangered species are publicity and education. By bringing the plight of endangered animals to more people, and by understanding how these creatures live, we can all help to save them.

African elephant

Protection

Most countries have laws protecting their endangered species, but sometimes these laws are difficult to enforce. One of the best ways to protect these animals is to set up national parks or wildlife refuges where hunting, logging and farming are not allowed. Some parks and refuges protect one particular species, while others preserve the natural habitat for all wild animals. Where the animals face a special threat, such as the illegal poaching of elephants and rhinoceroses, armed guards may be employed.

International effort

The Earth's wildlife is a heritage that belong to us all, but in the 20th century there was a huge, world-wide trade in wild animal parts – furs for fashion, ivory for jewellery, rhino horn for daggers, turtle shells for combs and spectacle frames, even tiger bones for use as medicine. An international agreement called CITES made this trade illegal, and tourists may now be searched for souvenirs made of endangered animal parts.

Hyacinth macaw

Back from the brink

To save endangered animals, we must understand how they behave in the wild. All over the world, scientists are researching animal behaviour. Some species are too rare in the wild even to count their numbers. In other cases, individual animals are captured and fitted with radio collars so their movements can be tracked. When a species becomes extinct in the wild, the only survivors are those already in captivity. If these animals breed, the species may be saved and eventually released into the wild. Lots of effort goes into starting 'captive-breeding' programmes before a species becomes extinct.

Quagga

Aa

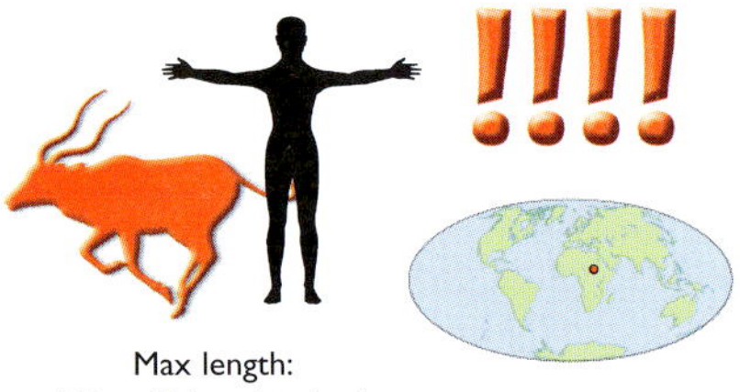

Max length:
1.7 m (5 feet 6 inches)

Addax

The addax is a very large desert antelope found only in the remotest parts of the southwestern Sahara Desert. Small herds of addax once roamed the whole desert, but now there are just a few – about 200 animals in total. The addax has been hunted for its meat and skin. Because of its size, this antelope is not capable of great speed.

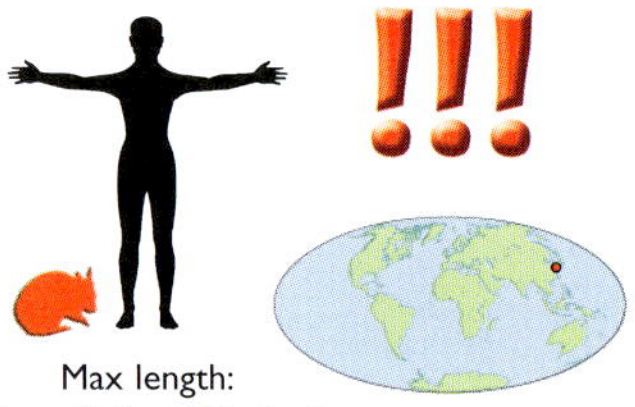

Max length:
51 cm (1 foot 7 inches)

Amami rabbit

With dark colouring and tiny ears, the Amami rabbit is found only in the mountain forests of two small islands off southern Japan. Habitat destruction has reduced the area of its natural habitat and only about 2,000 remain. The Japanese now protect this rabbit, and have given it the special status of a living 'natural monument'.

Max length: 5.8 m (19 feet)

American crocodile

The American crocodile is a saltwater species, found in mangrove swamps and coastal canals. This large reptile has been hunted for its tough hide, from which boots, bags and belts are made. It is also threatened by coastline development projects that are destroying its habitat. Now protected, the American crocodile's numbers are steadily increasing in Florida, although this is not the case in Haiti and El Salvador.

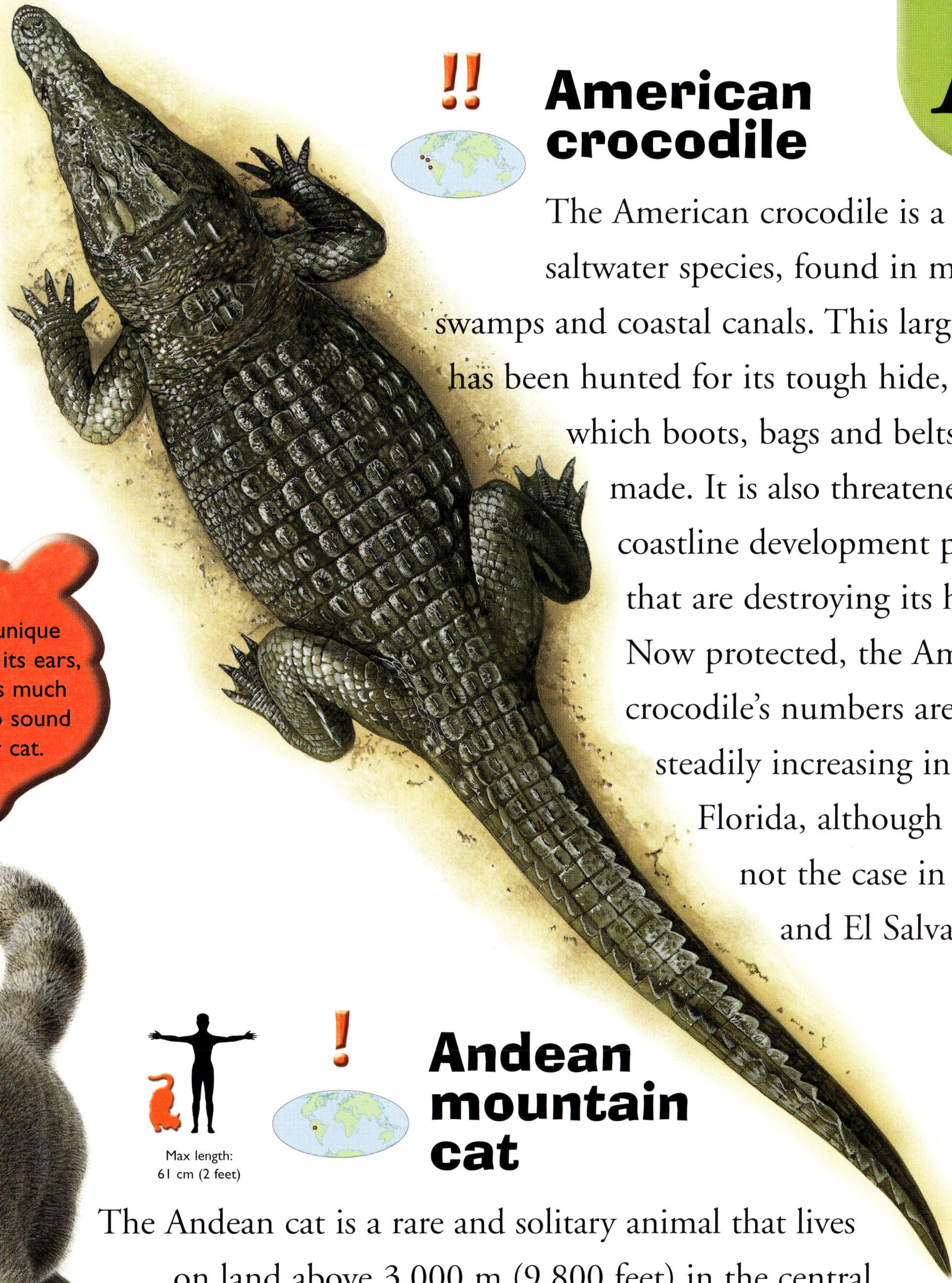

Fact

Because of the unique bone structure of its ears, the Andean cat is much more sensitive to sound than any other cat.

Max length: 61 cm (2 feet)

Andean mountain cat

The Andean cat is a rare and solitary animal that lives on land above 3,000 m (9,800 feet) in the central Andes. Very little is known about this small wild cat, and it is not directly threatened by human activity. However, it is considered endangered due to its rarity rather than because of hunting or habitat destruction.

Aa

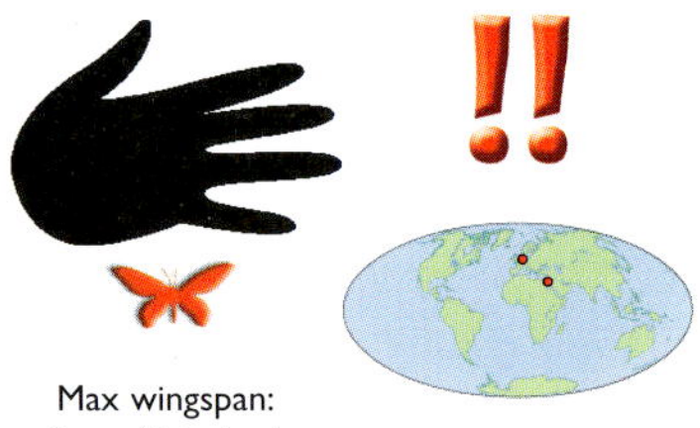

Max wingspan:
8 cm (3 inches)

Apollo butterfly

The Apollo butterfly has many subspecies around the world, and some European subspecies are showing an alarming decline in numbers. This is mainly caused by habitat destruction, air pollution affecting the insect's food plants, and butterfly collectors. The Apollo butterfly is also more vulnerable to predators as it spends two years as a caterpillar.

Fact
The aye-aye uses its especially long third finger to pull insects from their holes in wood.

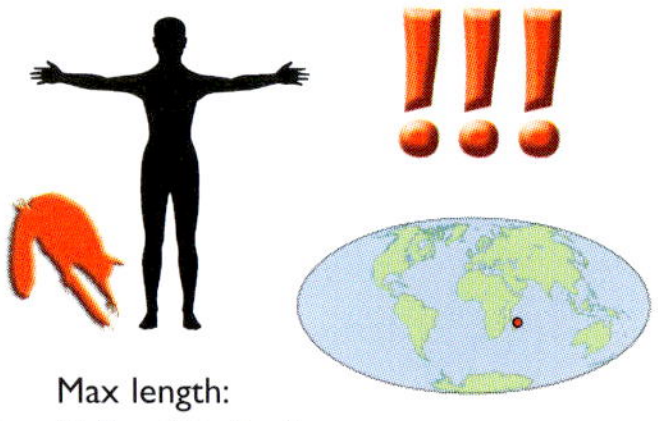

Max length:
1 m (3 feet 3 inches)

Aye-aye

The aye-aye is a small primate, related to lemurs, that hunts by night for wood-boring insects in the forests of Madagascar. The exact number of aye-ayes is unknown but it is believed to be between 1,000 and 2,000. Captive-breeding programmes have been established and some aye-ayes have been released back into the wild.

Max length:
3 m (10 feet)

!!!

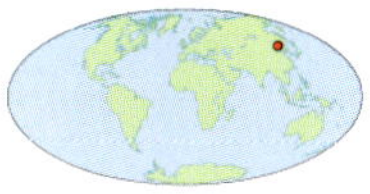

Bactrian camel

The Bactrian camel, with two humps, is found living in the cold, remote deserts of northern China and Mongolia. It is taller and leaner than the domesticated Bactrian camel and has smaller humps. The wild camel is endangered due to competition for water with domestic herds, and there are now fewer than 2,000 individuals.

Max length:
1 foot 1 inch (33 cm)

!!!

Bandicoot

The eastern barred bandicoot is a small marsupial, now almost extinct on mainland Australia, where only five naturally occurring animals remain. The captive-breeding programme has introduced 2,000 animals back into the wild. The island of Tasmania is the last stronghold of this shy, nocturnal animal. Here, the eastern barred bandicoot is threatened by habitat destruction and domesticated predators, such as cats and dogs.

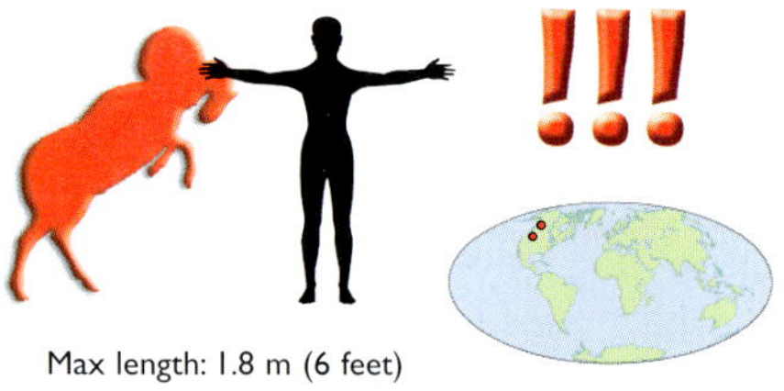

Max length: 1.8 m (6 feet)

Bighorn sheep

The bighorn sheep is found in the mountain pastures and rocky cliffs of the Rockies. Superbly adapted to life among almost-sheer rock faces, this sheep is agile enough to escape most of its natural predators, including the cougar (mountain lion). The biggest threat to the bighorn sheep is humans, both through habitat destruction and sport hunting. The Black Hills subspecies of the bighorn sheep is already extinct.

Max length:
3.7 m (12 feet 3 inches)

Black rhinoceros

The black rhinoceros was once widespread south of the Sahara. Now, fewer than 2,000 remain. It has been hunted almost to extinction for its horns, which are prized in the Middle East and China. For one horn, a poacher can earn twice what he would earn in a year as a farmer. The black rhino is now found only in heavily-guarded nature reserves, but illegal hunting still continues.

Max length: 91 cm (3 feet)

Black soft-shelled turtle

The black soft-shelled turtle is found only in one pond, next to a religious shrine in Bangladesh. The last 400 turtles live on food given to them by visitors. To save this unique species, scientists are trying to protect the turtles from disease.

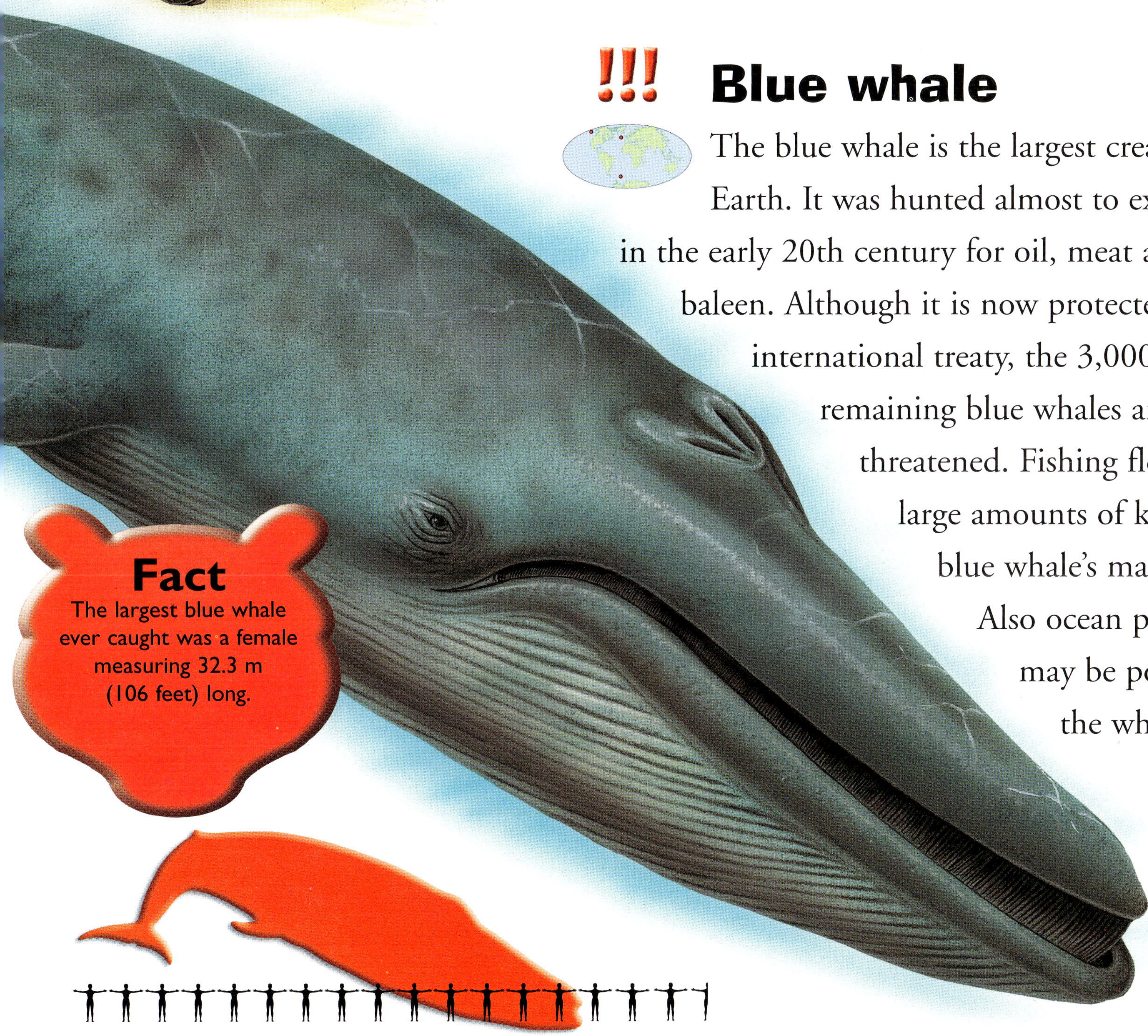

Blue whale

The blue whale is the largest creature on Earth. It was hunted almost to extinction in the early 20th century for oil, meat and baleen. Although it is now protected by international treaty, the 3,000 or so remaining blue whales are still threatened. Fishing fleets take large amounts of krill, the blue whale's main food. Also ocean pollution may be poisoning the whales.

Fact

The largest blue whale ever caught was a female measuring 32.3 m (106 feet) long.

Max length: 32.3 m (106 feet)

Cc

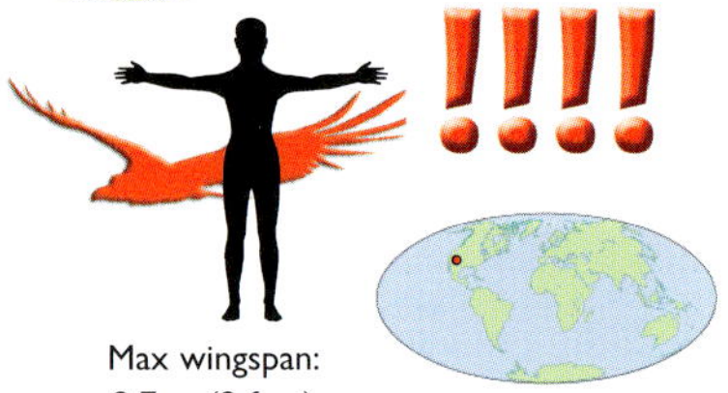

California condor

The California condor is one of the rarest birds in the world. In 1982, there were fewer than 25 left in the wild. Today, due to captive-breeding programmes, its numbers have increased to 200. The captive birds were released back into the wild in the 1990s. They will be recaptured if they fail to thrive. The condor became endangered through hunting, poisoning by pesticides and lead, and low reproduction rates.

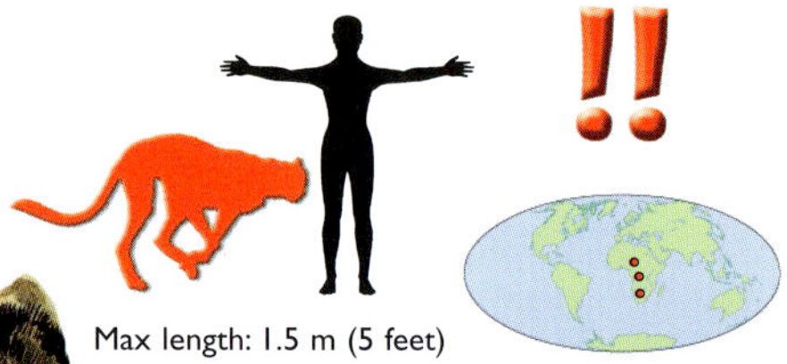

Cheetah

The cheetah is a magnificent predator and one of the fastest land animals. This large cat was once found throughout Africa, the Middle East, Asia and central India, but is now virtually extinct outside Africa, where about 10,000 remain. Habitat destruction, fur hunters and disease have caused the drop in numbers. Cheetahs are now protected by law and captive-breeding programmes are trying to improve their resistance to disease.

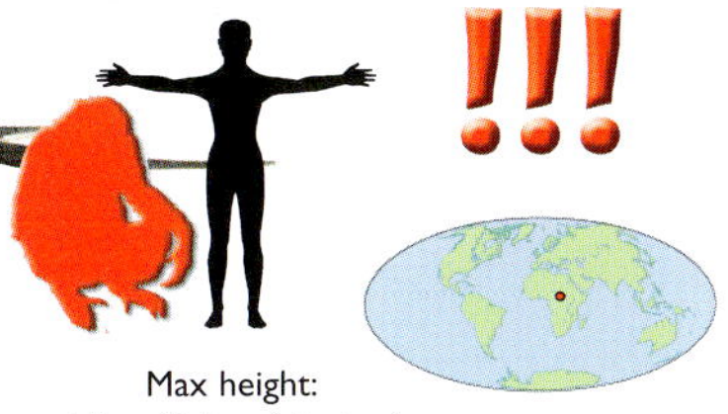

Max height:
1.7 m (5 feet 6 inches)

Chimpanzee

The chimpanzee is our closest living relative and is found in the tropical forests of Africa. There are three varieties, the common, masked, and long-haired, all of which are now endangered because of human interference. The chimps face habitat destruction, hunting, and being captured as pets, or for zoos and research laboratories.

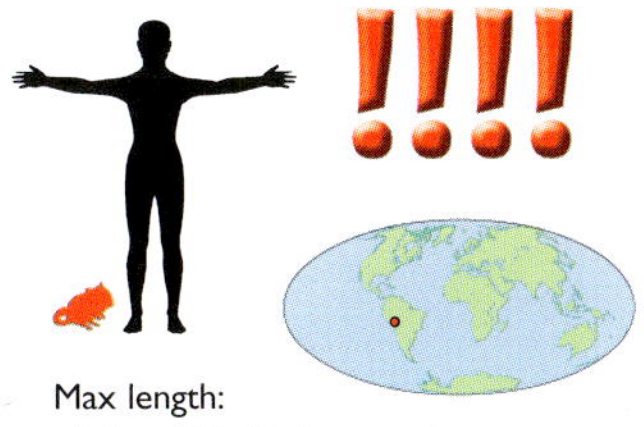

Max length:
38 cm (1 foot 3 inches)

Chinchilla

The long-tailed chinchilla is a small rodent with thick, soft fur that lives in the barren, arid areas of the Andes mountains. By the 1940s, fur hunters nearly caused its extinction. Since then, wild populations are slowly recovering due to legal protection. However, the short-tailed chinchilla is critically endangered and remains on the verge of extinction.

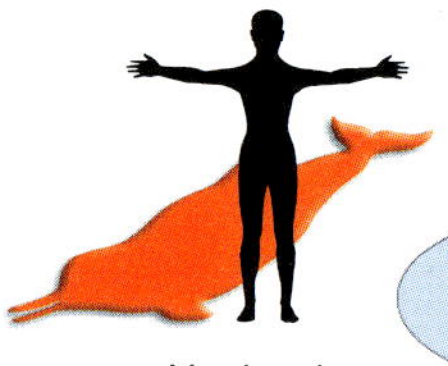

Max length:
2.5 m (8 feet 3 inches)

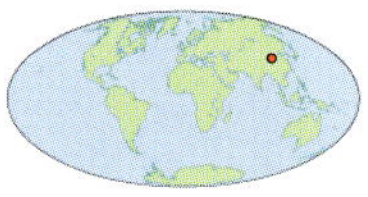

!!!! Chinese river dolphin

The Chinese river dolphin, also known as the baiji, is found only in the lower part of the Yangtze River. It is now critically endangered because of overfishing, water pollution, collisions with boats, and the building of dams that interfere with its migration. In 1984, there were approximately 400 individuals in the wild. Today, there are only five, and one in captivity.

Max length:
1.8 m (5 feet 11 inches)

!!!! Coelacanth

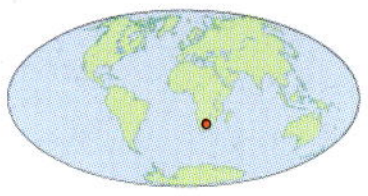

The East African coelacanth is one of two fish species (the other is the Indonesian coelacanth) that are 'living fossils'. Until its discovery in the 1930s, the coelacanth was thought to have died out some 65 million years ago. It is a very rare creature, endangered because it sometimes gets caught in fishing nets intended for other species.

!!! Crested gibbon

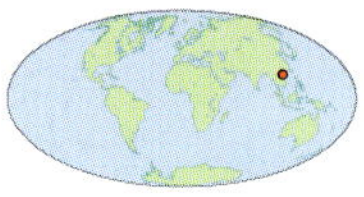

The crested gibbon is found only in the forested mountains of northern Vietnam and southern China. It lives in the trees, swinging from the branches on arms that are longer than its legs. Habitat destruction, meat hunting and its use in traditional medicines have made it the most endangered of all the apes, with fewer than 50 left in the wild.

Max height:
64 cm (2 feet 1 inch)

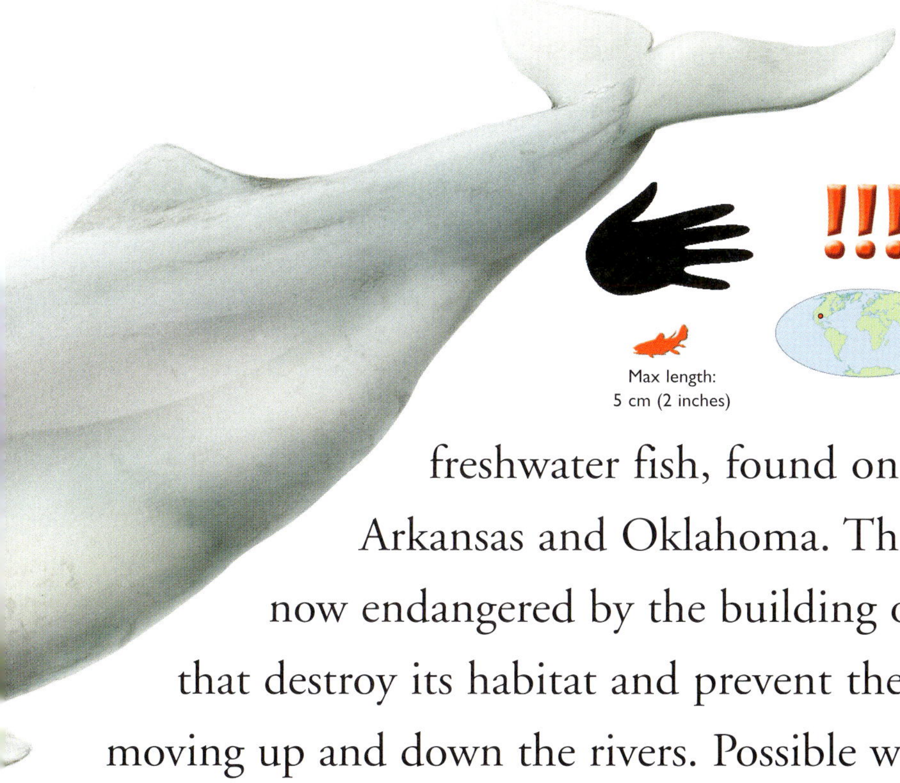

Max length:
5 cm (2 inches)

!!! Darter

The leopard darter is a small freshwater fish, found only in a few rivers in Arkansas and Oklahoma. This rare fish is now endangered by the building of reservoirs that destroy its habitat and prevent the fish from moving up and down the rivers. Possible ways of saving the leopard darter are now being studied.

Fact
More than 10,000 desmans have been bred in captivity and then released into areas with little water pollution.

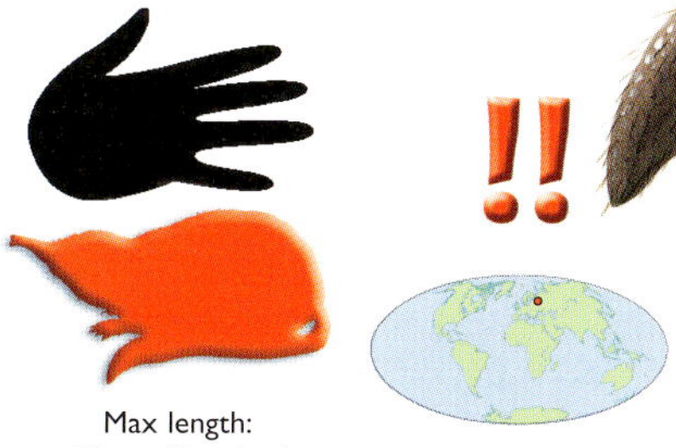

Max length:
23 cm (9 inches)

!! Desman

The Russian desman is a relative of the mole family that was hunted almost to extinction for its thick, soft fur. It spends most of its time in rivers and streams searching for crayfish and amphibians. Animals bred in captivity have been released in some areas. However, water pollution and habitat destruction have limited the efforts to reintroduce the desman into its original range.

Dd

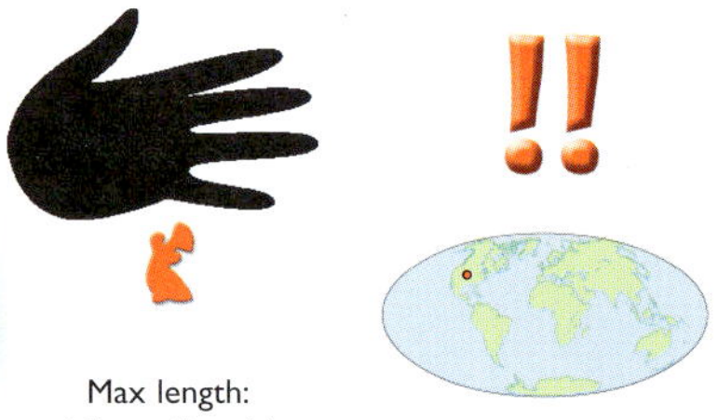

Max length:
2.5 cm (1 inch)

Devil's Hole pupfish

The Devil's Hole pupfish is found in one small desert spring in the Mojave desert, where it has been for 10-20 thousand years. The entire population ranges from 100-500 fish, depending on the season. If the water level drops, or any link in the chain is broken, the pupfish will become extinct. To safeguard against this, scientists have built two special ponds, each with 200 fish.

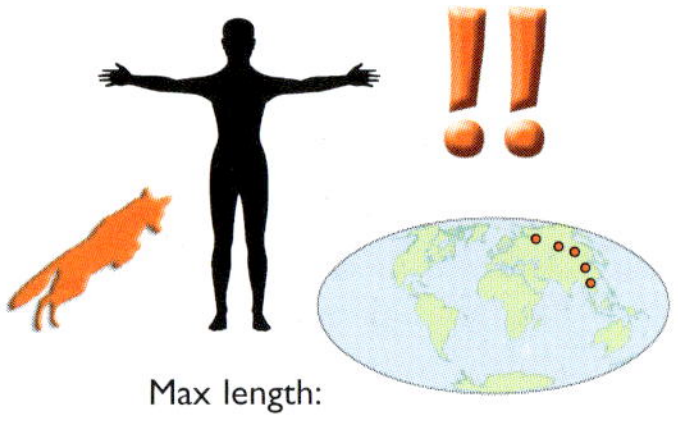

Max length:
91 cm (3 feet)

Dhole

The dhole, or Asian red dog, lives and hunts in small packs. Once widespread, its numbers and range have declined dramatically – mainly because large areas of its habitat have been cleared to make way for crops. Throughout history, the dhole was viewed as a pest, causing it to be poisoned or shot.

Fact
An adult dhole can jump vertically more than 2.3 m (7 feet 6 inches) into the air.

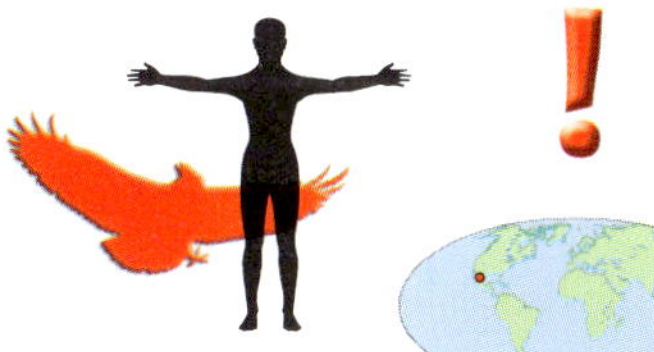

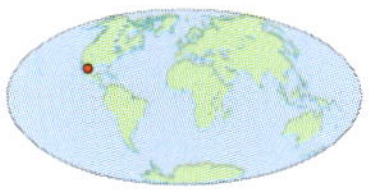

Max wingspan:
2 m (6 feet 6 inches)

Eagle, harpy

The harpy eagle is a powerful rainforest predator of Central and South America with a wingspan of more than 1.8 m (6 feet). It hunts over a large area. Its numbers are declining because of habitat destruction as the rainforest is cut down. The patches of forest left are not big enough for the eagles, scattering their population.

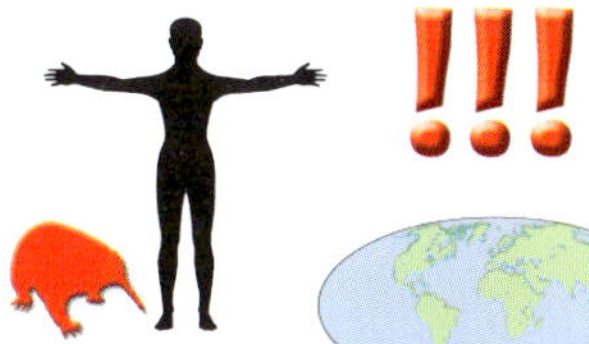

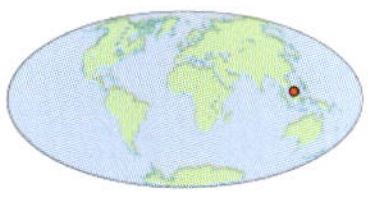

Max length:
76 cm (2 feet 6 inches)

Echidna

The rare long-nosed echidna lives in the mountain forests of New Guinea. With the short-nosed echidna and the platypus, it is one of only three egg-laying mammals. It moves slowly, so it is often hit by cars. Also, logging is threatening its habitat.

Ee

Max height:
3.3 m (10 feet 10 inches)

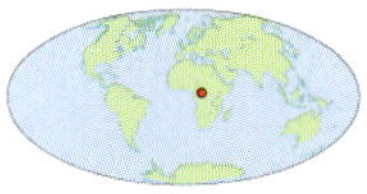

!!! Elephant (African)

The African elephant is the largest land mammal. It has distinctively large ears and tusks, for which it is hunted. There are approximately 400-600 thousand remaining and, despite armed guards and global treaties, illegal poaching still threatens these magnificent animals.

Max height: 3 m (10 feet)

!!! Elephant (Asian)

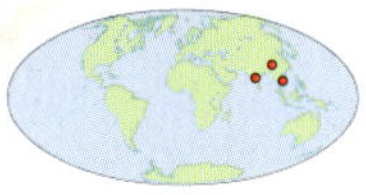

The Asian elephant is smaller than its African cousin, and can be easily domesticated as a working animal. It is sometimes (incorrectly) called the Indian elephant. There are about 35-50 thousand Asian elephants still living in the wild, but they are endangered by the destruction of their habitat and also by hunting. Several countries have established wildlife refuges in an effort to protect these elephants.

Max wingspan:
80 cm (2 feet 7 inches)

Falcon, peregrine

The peregrine falcon is a very fast-flying bird of prey that is found in most parts of the world. In the middle of the 20th century, the number of peregrine falcons began declining dramatically, especially in Europe and North America. The birds were being affected by DDT, a pesticide that was widely used. Thanks to strict controls on the use of DDT, populations of these birds are starting to recover.

Fact
The peregrin falcon is the fastest bird in the world, reaching speeds of up to 350 km/h (217 mph) as it dives through the sky.

Max length:
60 cm (2 feet)

Ferret, black-footed

The black-footed ferret is a small mammal, once found throughout the Rockies and Great Plains, where it preyed on prairie dogs. When the prairie dogs were nearly wiped out because farmers saw them as pests, the wild ferret was thought to be extinct. In 1981, a small population was discovered in Wyoming. This population plus the captive-breeding programme has saved the species.

Ff

Max length:
1 m (3 feet 3 inches)

!!!

Flightless cormorant

The flightless cormorant is found only on two of the Galapagos Islands. This rare species dives from coastal rocks to hunt fish underwater. Before people colonized the islands, the cormorant was free of land predators. Its numbers have been drastically reduced since human settlers have introduced cats, rats and pigs.

Fact
The flightless cormorant does not have oily feathers. Air trapped between the feathers prevents the cormorant from becoming waterlogged.

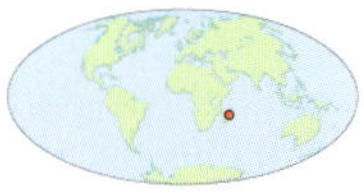

Max length:
80 cm (2 feet 8 inches)

!!!

Fossa

The fossa is a medium-sized predator of Madagascar, related to the mongoose. Its natural rainforest habitat and prey (mainly lemurs) have both been greatly reduced by human activity. As a result, fossas began raiding farms and taking chickens. The species is widely hunted and its habitat is being encroached upon.

Max length:
2.5 m (8 feet 3 inches)

Ganges dolphin

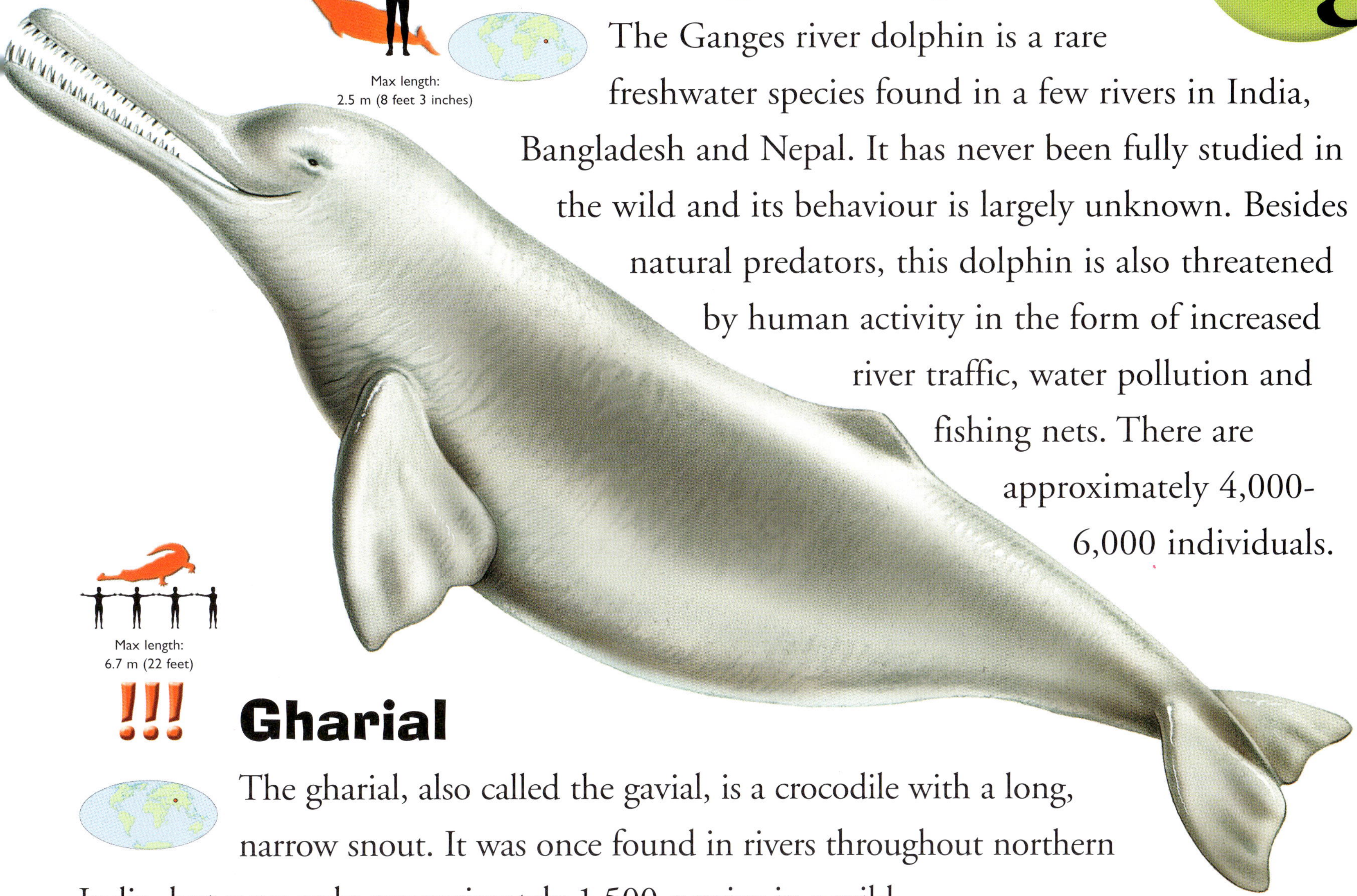

The Ganges river dolphin is a rare freshwater species found in a few rivers in India, Bangladesh and Nepal. It has never been fully studied in the wild and its behaviour is largely unknown. Besides natural predators, this dolphin is also threatened by human activity in the form of increased river traffic, water pollution and fishing nets. There are approximately 4,000-6,000 individuals.

Max length:
6.7 m (22 feet)

Gharial

The gharial, also called the gavial, is a crocodile with a long, narrow snout. It was once found in rivers throughout northern India, but now only approximately 1,500 survive in a wild population. This is because of habitat destruction and being hunted for meat and skin. The gharial's eggs and body parts are also prized for use in traditional medicine. Conservation and captive-breeding programmes have helped somewhat. Over 3,000 animals have been released back into the wild.

Gg

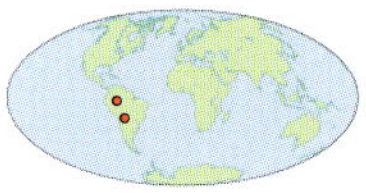

Max length:
1 m (3 feet 3 inches)

!!!

Giant armadillo

The South American giant armadillo is a mammal covered in armour-plating made of horn and bone. It lives alone, sheltering by day in burrows dug with its huge claws. At night it feeds mainly on insects. Habitat loss, especially through logging, has caused numbers to fall to just a few thousand animals scattered across the continent.

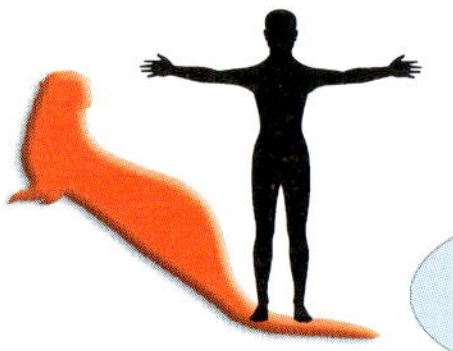
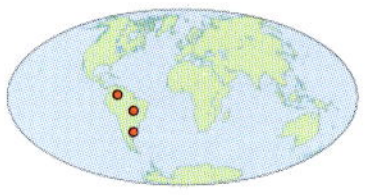

Max length: 1.5 m (5 feet)

!!!

Giant otter

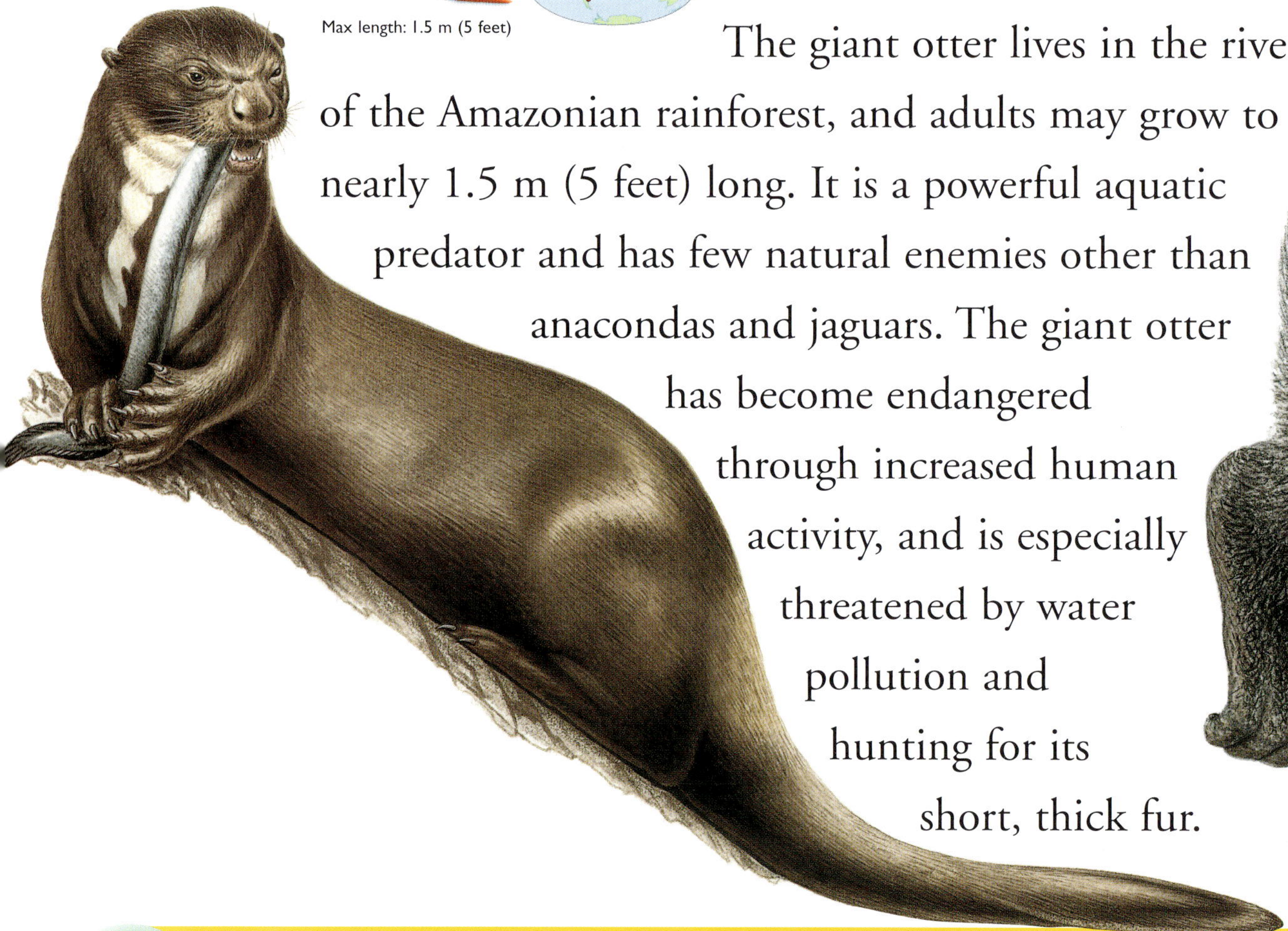

The giant otter lives in the rivers of the Amazonian rainforest, and adults may grow to nearly 1.5 m (5 feet) long. It is a powerful aquatic predator and has few natural enemies other than anacondas and jaguars. The giant otter has become endangered through increased human activity, and is especially threatened by water pollution and hunting for its short, thick fur.

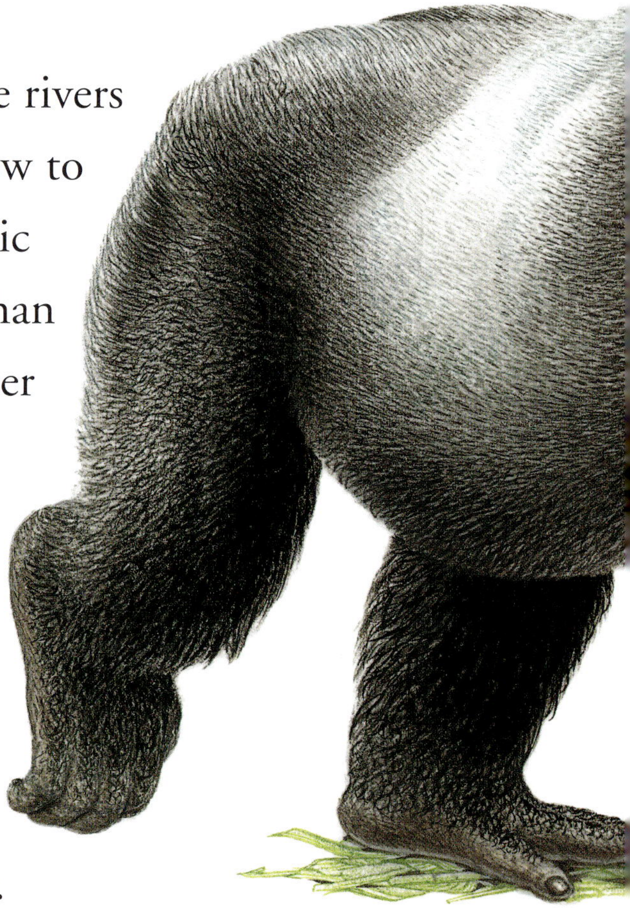

Gg

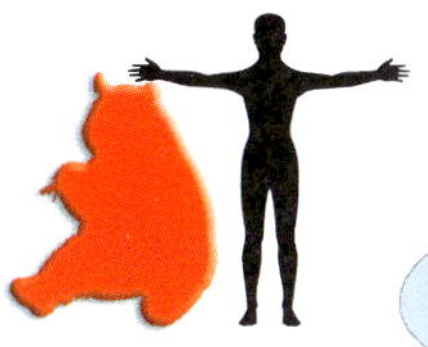
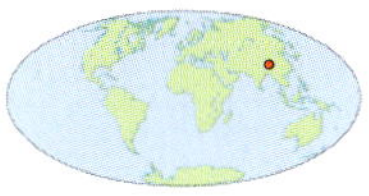

Max length:
1.5 m (5 feet)

Giant panda

The giant panda has always been a rare animal, confined to the mountain forests of central China, where it eats primarily bamboo. This bear-like mammal has been hunted to near extinction for its fur, and has also suffered from habitat destruction. Despite laws to protect them, only about 1,000 giant pandas remain in the wild.

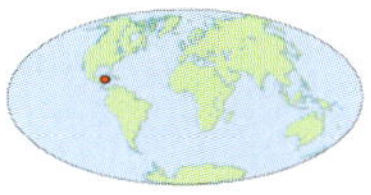

Max length: 5 cm (2 inches)

Golden frog

The golden frog is found only in isolated regions of Panama. Its bright colour warns predators that it is toxic. Scientists believe that a major cause of its decline is climate change. During drought years, the frogs are forced into overcrowded wet areas, which lead to fatal diseases.

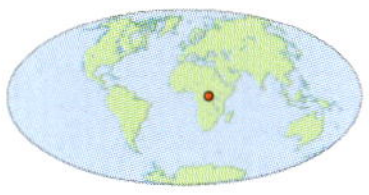

Max height:
1.9 m (6 feet 3 inches)

Gorilla

The mountain gorilla is the largest primate, weighing 182-226 kg (400-500 pounds) and it is also the most endangered. Fewer than 600 now live in two isolated populations, in a small area of cloud forest. Strict laws exist to protect the gorilla, but local war has threatened its protected status as its habitat is threatened and poaching continues.

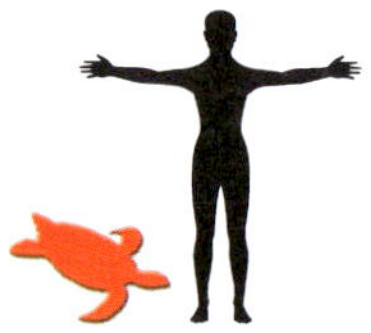

Max length:
1 m (3 feet 3 inches)

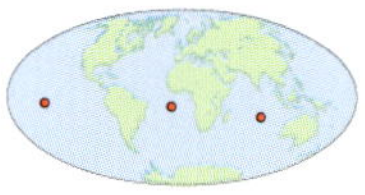

Hawksbill turtle

The hawksbill turtle, always hunted for its attractive 'tortoiseshell', seems headed for extinction. The main reason it faces extinction is the increasing number of eggs stolen. More than half of the eggs laid by a hawksbill turtle are stolen by humans. The turtle's coral reef feeding grounds are in danger also. It may be too late to save this species.

Fact

The hawksbill is the only sea turtle with a serrated edge around its shell.

Max wingspan:
3.3 cm (1.3 inches)

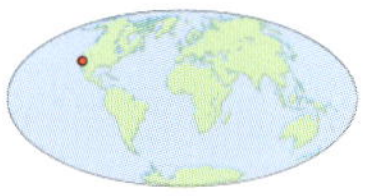

Hermes copper butterfly

The Hermes copper butterfly is a rare species, living only in one tiny area near San Diego, California. Its numbers are declining because of habitat destruction and fires. Efforts are being taken to protect the existing colonies from habitat loss and separatation from each other.

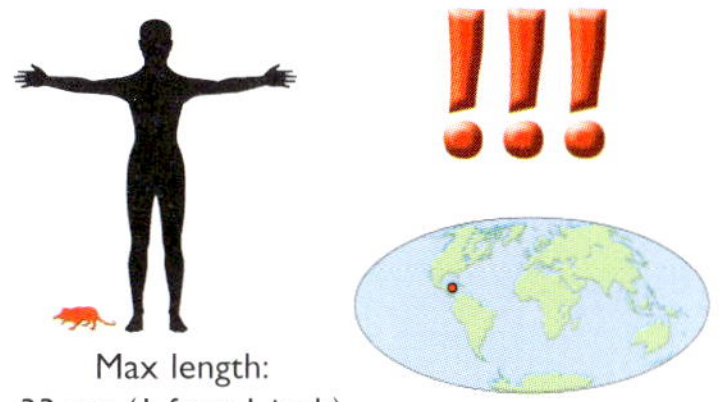

Max length:
33 cm (1 foot 1 inch)

Hispaniolan solenodon

The Hispaniolan solenodon is a small, insect-eating mammal. It is found only on the Caribbean island of Hispaniola, but a close cousin lives on the nearby island of Cuba. Both species are endangered because large numbers of solenodons have been killed by the cats and dogs introduced by human settlers.

Max length: 60 cm (2 feet)

Hispid hare

The hispid hare, or bristly rabbit, one of the world's rarest mammals, was once thought to be extinct. Today, just 100 hispid hares are living wild in the southeastern Himalayan foothills. Habitat destruction and hunting continue to threaten the future of the hispid hare.

Max length:
1 m (3 feet 3 inches)

Hyacinth macaw

The hyacinth macaw, largest of all parrots, is also one of the most endangered, so popular is it as an attractive cage bird. Despite its protected status, large numbers are being trapped and sold illegally each year. It is estimated that there are only 2,500 left in the wild.

Max length:
96 cm (3 feet 2 inches)

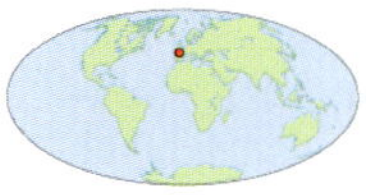

Iberian lynx

The Iberian lynx is now found only in a few mountainous areas of Spain and Portugal, where it feeds on rabbits and ducks. This small cat has been hunted almost to extinction for its beautiful, spotted fur. The Iberian lynx is now protected by law, but remains one of the world's most endangered cats.

Max length:
1.7 m (5 feet 6 inches)

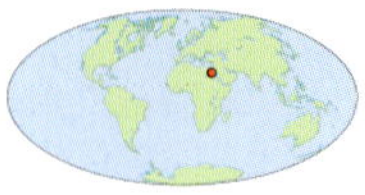

Ibex

The ibex is a mountain goat with very large, thick, curved horns. It has a wide distribution but its numbers are declining rapidly because of intensive hunting. The estimated 1,200 ibex that survive are found on the highest and most inaccessible areas of mountain pasture.

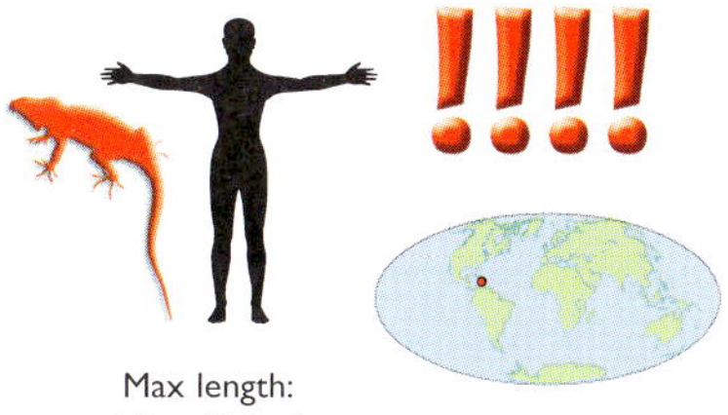

Max length:
1.7 m (5 feet)

Iguana, Grand Cayman blue

The Grand Cayman blue iguana is one of the most endangered reptiles. It can be found only in one small part of a tiny Caribbean island. These iguanas nest on the open rock surface, so they are easy prey for domesticated cats and dogs. The total population of Grand Cayman blue iguanas is between 100-200 animals. Without intervention, the surviving wild population will be extinct within the next five years.

Fact
Local people know the indri as a babakato, which means 'little father of the forest'.

Max length: 91 cm (3 feet)

Indri

The indri, or babakoto, is the largest Madagascan lemur. It travels in giant leaps or climbs trees vertically. Local taboos forbid hunting the indri, so its biggest threat is habitat destruction. Increased forest clearing is the result of Madagascar's exploding human population, and most of the trees that provide leaves for the indris to eat have been cut down.

Jj

Max length: 1.8 m (6 feet)

Jaguar

The jaguar is found from the southwestern US to Central South America. It has been hunted for its fur since ancient times. The jaguar is threatened by the destruction of its habitat as this big cat requires a large territory to provide enough prey. Also, despite the ban on hunting, poaching is still a significant threat to the jaguar.

Japanese crane

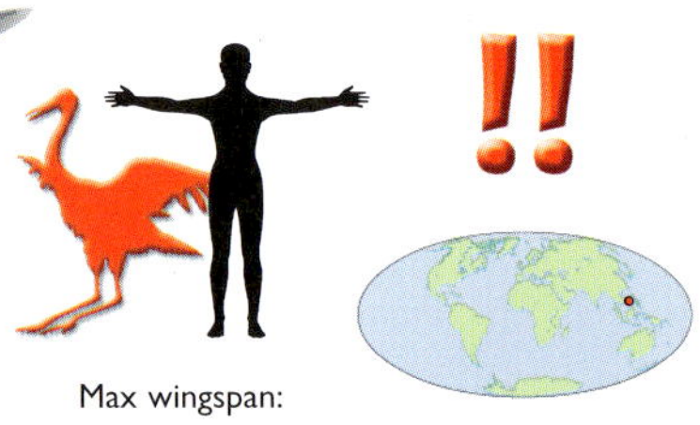

Max wingspan: 2.7 m (9 feet)

The Japanese crane, also known as the red-crowned crane, famed for its elaborate courtship dance, is a symbol of long life. Even so, large numbers of these birds were killed as they migrated between Siberia and Japan. By the beginning of the 20th century, fewer than 50 birds survived. Since then, the Japanese crane has been protected by law and its numbers have risen to approximately 1,500 birds. Habitat destruction still endangers the remaining cranes.

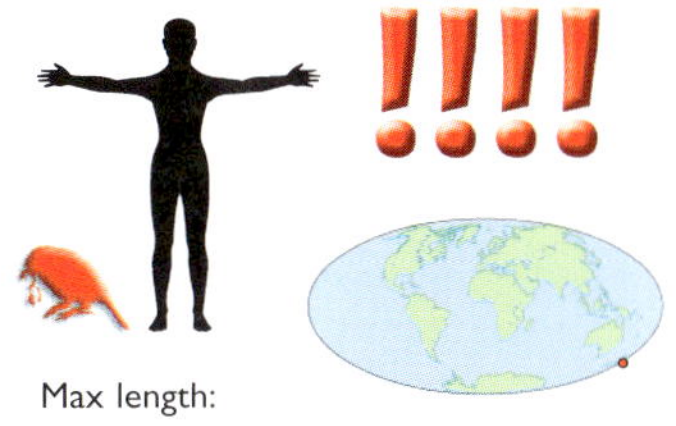

Max length:
64 cm (2 feet 1 inch)

Kakapo

The kakapo is a flightless parrot that lives in New Zealand, where it had few natural predators. However, the cats, rats and dogs introduced by human settlers resulted in a massive decline in the kakapo population. The species has been saved by a captive-breeding programme. The remaining 62 birds have been released on three remote, predator-free islands in an attempt to re-establish wild populations.

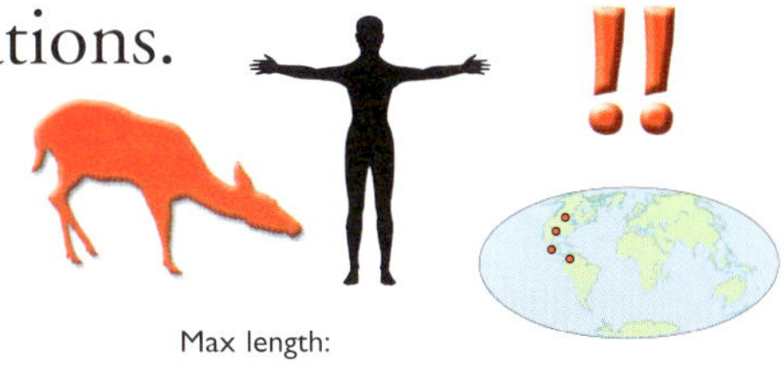

Max length:
1.9 m (6.5 feet)

Key deer

The Key deer is found only on a few of the islands of the Florida Keys. Hunting and habitat destruction in the early 20th century left just 50 deer. Since then, Key deer have been protected in a wildlife refuge. Numbers have risen steadily and about 300 now live in the wild.

Max length:
76 cm (2 feet 6 inches)

Kinkajou

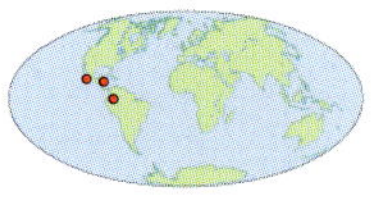

The kinkajou, or martucha, is related to the raccoon and panda. Its strong, prehensile tail makes it an excellent tree climber. It feeds by night on fruit, insects and honey, sharing the nickname of 'honey bear' with the endangered sun bear. The clearing of its rainforest home is endangering the kinkajou.

Max length:
40 cm (1 foot 4 inches)

Kiwi

The kiwi is a flightless bird that lives only in New Zealand, where it is the unofficial national symbol. Once hunted for its feathers, it is now protected – but still at risk from cats and rats introduced by humans, and by the clearing of its woodland habitat.

Max length:
11 cm (4.4 inches)

Knysna seahorse

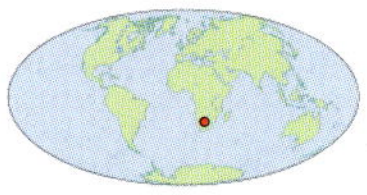

The Knysna seahorse is found only in a few estuaries and lagoons near the Cape of Good Hope in South Africa. As well as its high sensitivity to pollution, it is also threatened by the demand for seahorses in traditional medicine. Each year, tonnes of these tiny creatures are caught, leaving the population depleted.

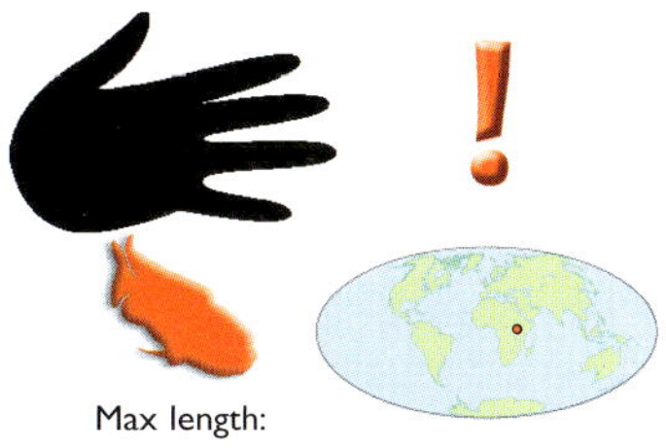

Max length:
10 cm (4 inches)

Lake Victoria cichlid

Lake Victoria once had hundreds of species of cichlid fish that were found only in that lake. *Haplochromis nuschisquamulatus* (pictured) is the scientific name of just one of these. In the 20th century, the Nile perch was introduced to the lake as a food fish to boost the local industry. Unfortunately, the Nile perch has a huge appetite that has wiped out many of Lake Victoria's cichlid species.

Fact
With some cichlids, newly hatched young are raised inside the mouth of the male parent.

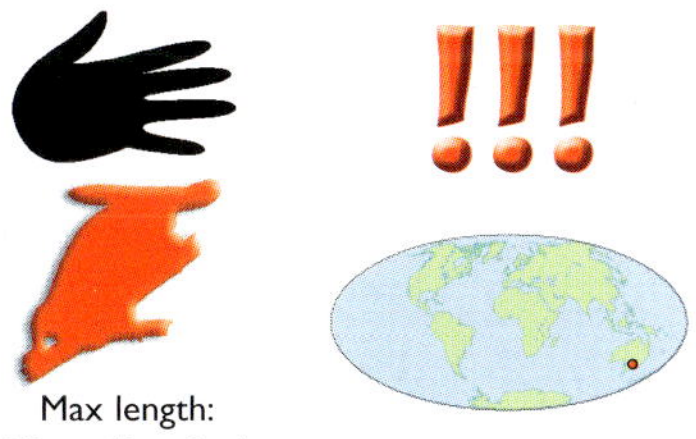

Max length:
15 cm (6 inches)

Leadbeater's possum

Leadbeater's possum is a small, tree-dwelling marsupial that was declared extinct in 1909. Fifty years later it was discovered to be living in remote highland areas of Victoria, Australia. Thanks to conservation and protection measures, Leadbeater's possum appears to be flourishing once more, but its numbers are still dangerously low.

Ll

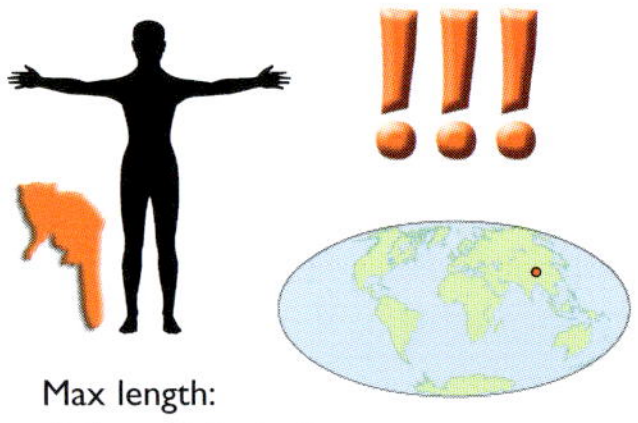

Max length:
58 cm (1 foot 11 inches)

Lesser panda

The lesser panda, or red panda, is related to the raccoon. It lives in the forests of the Himalayan foothills. Like the giant panda, a distant relative, it mainly eats bamboo leaves. The lesser panda is endangered by environmental changes. However, it seems to be benefiting from the protection of the giant panda. Some lesser pandas bred in captivity have been released back into the wild.

Fact
The lesser panda has partially retractable claws that make it an excellent tree climber.

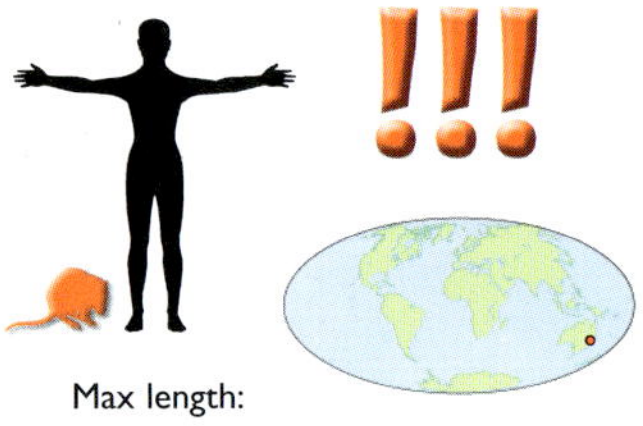

Max length:
43 cm (1 foot 5 inches)

Long-footed potoroo

The long-footed potoroo, which is also known as the rat kangaroo, is found only in the wetter forested areas of southeastern Australia. This small marsupial is active at night and feeds mainly on fungi that it digs up with its strong front limbs. The long-footed potoroo is threatened by habitat destruction and introduced predators, including the red fox, dingo and feral dog.

Max length:
4.5 m (15 feet)

!!!

Manatee, West Indian

The West Indian manatee is a harmless sea mammal that lives in rivers, bays, canals and coastal waters (fresh or salt water), where it feeds exclusively on vegetation. It is large and slow-moving – an adult can weigh up to 544 kg (1,200 pounds). The West Indian manatee is endangered by water pollution, loss of habitat, and hundreds are injured or killed by boat propellers. There are less than 2,000 left in the world.

Max length:
18 cm (7 inches)

!!

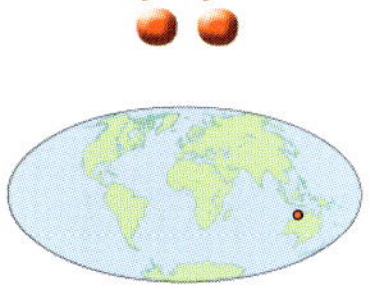

Marsupial mole, southern

The marsupial mole is not related to other moles. It lives in the deserts and scrublands of Australia. Its powerful limbs and large claws allow it to 'swim' through loose, sandy soil in search of lizards, worms and insects. Its eyes are very tiny and hidden. It is threatened by predators such as cats, dogs and dingoes.

Mm

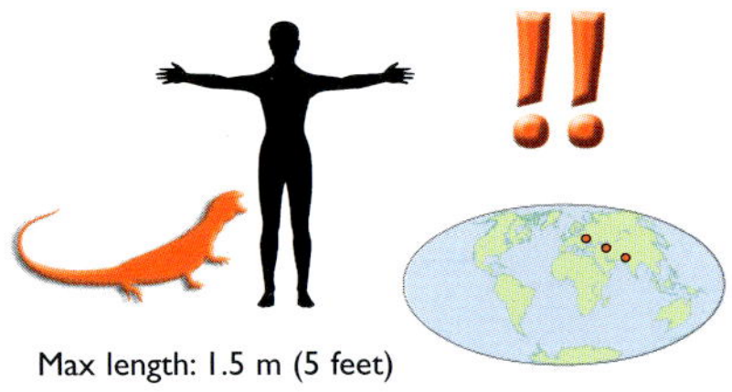
Max length: 1.5 m (5 feet)

Monitor

The Caspian desert monitor is the largest desert lizard in the world, reaching up to 1.5 m (5 feet) from the tip of its nose to the end of its flattened tail. It is one of the rarest lizards, having been hunted almost to extinction by desert travellers. The Caspian desert monitor is now protected by law in Russia. Its status in Afghanistan is uncertain.

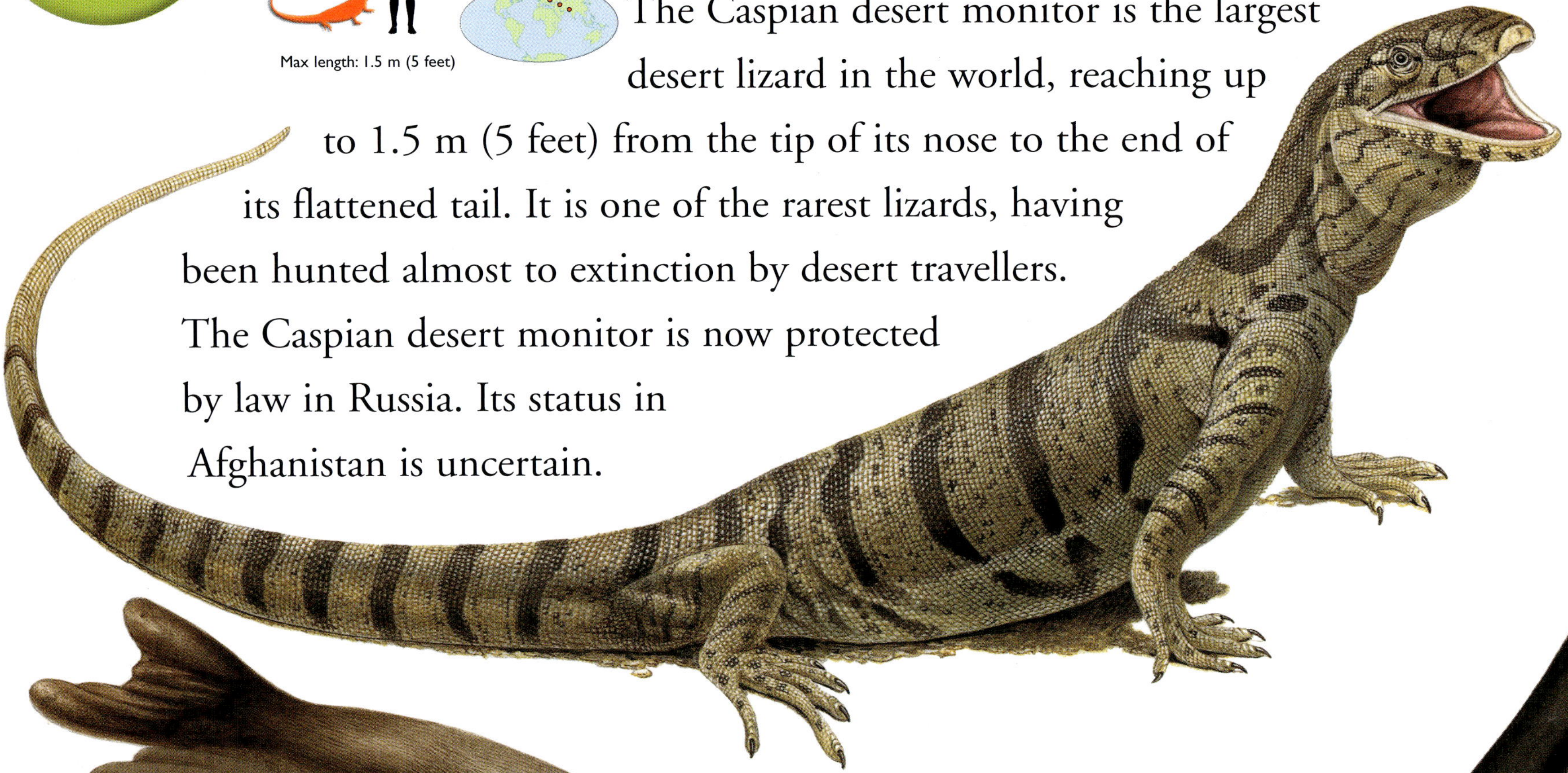

Max length:
2.5 m (8 feet 2 inches)

Monk seal

The Mediterranean monk seal is one of the world's rarest and most endangered mammals. Pollution, habitat destruction and overfishing are threatening this seal. Only 500 or so remain in the remotest areas of the Mediterranean. The Caribbean monk seal is already extinct.

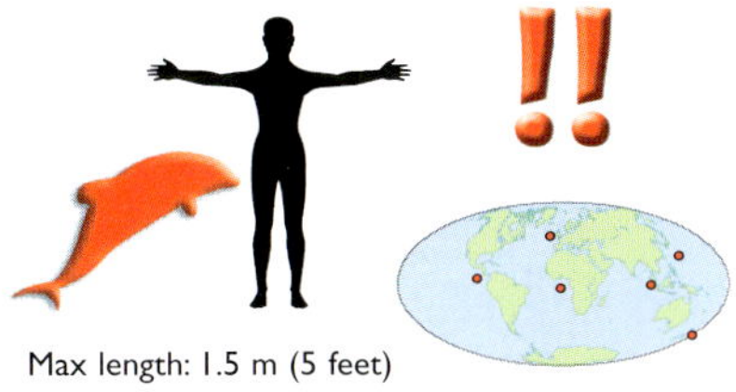
Max length: 1.5 m (5 feet)

New Zealand dolphin

The New Zealand dolphin is also known as Hector's dolphin. It looks more like a small porpoise than a dolphin and lacks the distinctive beak. The New Zealand dolphin lives in coastal waters and is now threatened by water pollution from coastline developments. Each year, many of these mammals become entangled in fishing nets and drown.

Fact
Sightings of the night parrot are so rare – about one every ten years – that they are reported in the newspapers.

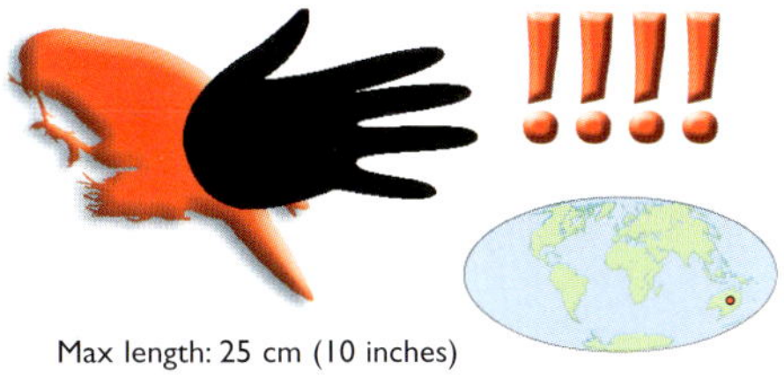
Max length: 25 cm (10 inches)

Night parrot

The night parrot is one of the world's rarest birds and it has been sighted only a few times during the last 100 years. Scientists have no idea how many of these elusive birds remain scattered across the Australian outback – it may be as few as 50 or less. The night parrot population has been drastically reduced by the spread of farming and by the introduction of predators such as cats and dogs.

Nn

Max length:
80 cm (2 feet 6 inches)

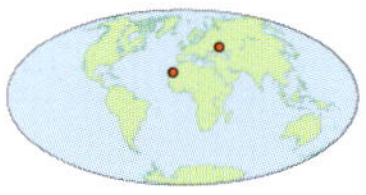

!!!!

Northern bald ibis

The northern bald ibis is now at the point of extinction with only 220 birds left in the world. This is the result of centuries of habitat destruction and pesticides. Once fairly common in southern Europe and North Africa, the northern bald ibis is now only found in three small colonies located in Morocco and Syria. Guards are posted in Syria to watch over the colony.

!!!!

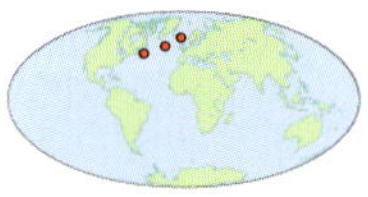

Northern right whale

The northern right whale is now the most endangered of all the large whales, despite being protected by law since the 1930s. Only about 300 of these whales remain in the oceans and their numbers are still declining. Some scientists believe that heavy shipping traffic and increased ocean pollution will bring about certain extinction.

Max length: 16.8 m (55 feet)

Fact

The right whale got its name because it was considered the 'right' whale to hunt for food and oil.

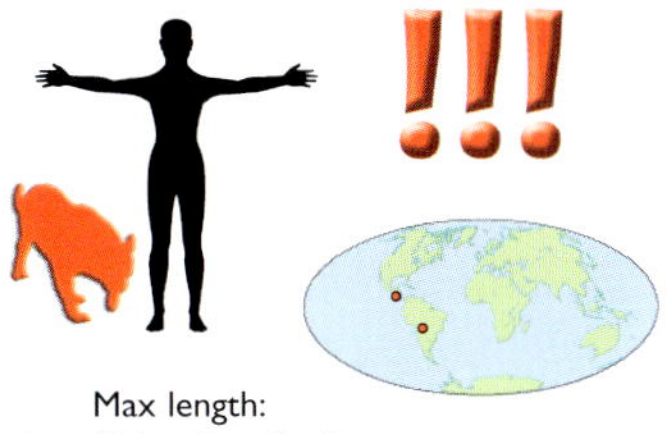
Max length:
1 m (3 feet 3 inches)

Ocelot

The ocelot is a medium-sized forest predator, living in areas of dense scrub to tropical rainforests. During the 20th century, millions of these cats were killed for their fur. The ocelot is now protected from hunters by law and international treaty, but it is still threatened by habitat destruction as forests are cut down for timber and farmland.

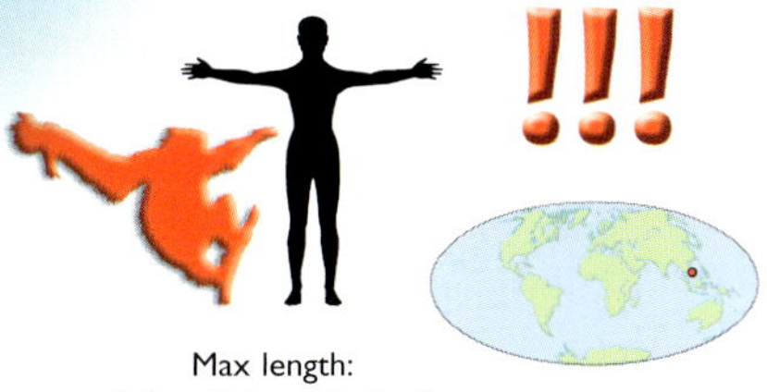
Max length:
1.4 m (4 feet 6 inches)

Orangutan

The orangutan is a smaller relative of the gorilla that is extinct on the mainland of southern Asia and found only on a few islands. It is threatened both by the destruction of its tropical rainforest habitat and by hunters seeking animals for the pet trade, despite its protection by law. Several forest sanctuaries have been established to protect the orangutan.

Oo

Max wingspan:
2.4 m (8 feet)

Oriental white stork

The oriental white stork is a large migratory bird. Unlike its European cousin, it has a black bill. Now highly endangered, only about 2,500 white storks still survive. In Japan, where the bird used to build its nests on the roofs of houses, it has been extinct since the 1970s. The main reason for its decline is the draining of its wetland breeding grounds in mainland Asia.

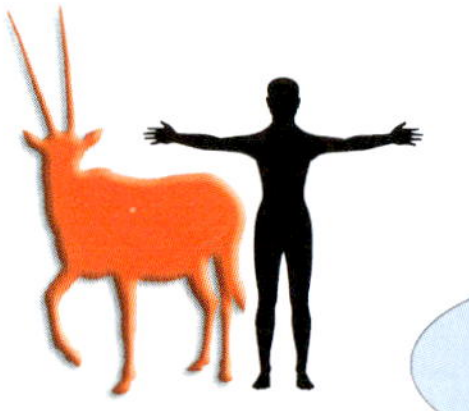

Max length:
1.7 m (5 feet 5 inches)

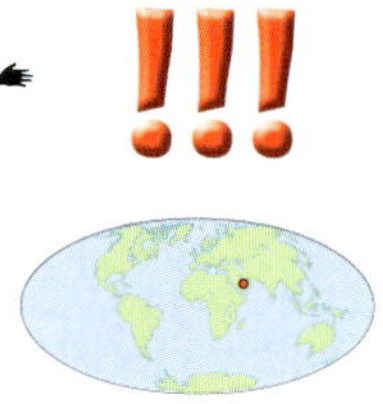

Oryx

The Arabian oryx is a desert antelope that has been rescued from near extinction by a captive-breeding programme that has released almost 100 back into the wild. Almost all the wild Arabian oryx were killed in the last century by hunters. A few were rescued and kept on a nature reserve in the Sultanate of Oman. As the size of the captive herd increases, more and more animals are released back into the wild.

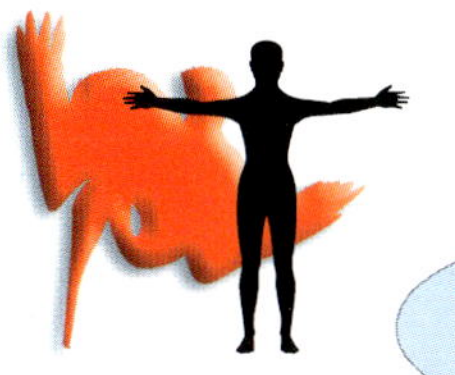
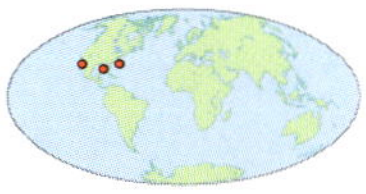

Max wingspan:
2.8 m (9 feet 6 inches)

!!

Pelican, brown

The brown pelican has now regained its former numbers across the Atlantic coast of the USA and the Gulf coast of Florida. It is still endangered in the rest of its range. In the 20th century, brown pelican populations were devastated by eating fish that contained DDT. The pesticide made the pelicans lay very fragile eggs that were easily damaged.

Fact
The Galapagos penguin is the world's smallest penguin and the only one that lives at the equator.

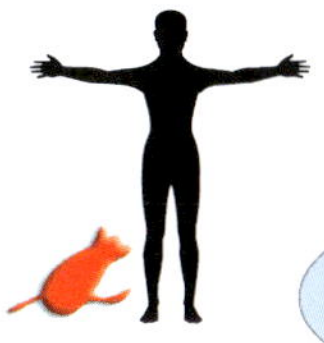
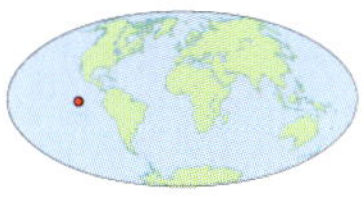

Max length:
53 cm (1 foot 9 inches)

!!!

Penguin, Galapagos

The Galapagos penguin is a rare species that lives on the remote tip of just one island. There are fewer than 1,500 of these seabirds left. Scientists have found that, in the years with warm El Niño currents, there was a sharp decline in the penguin population. The impact of predators on the Galapagos penguin population is also being investigated.

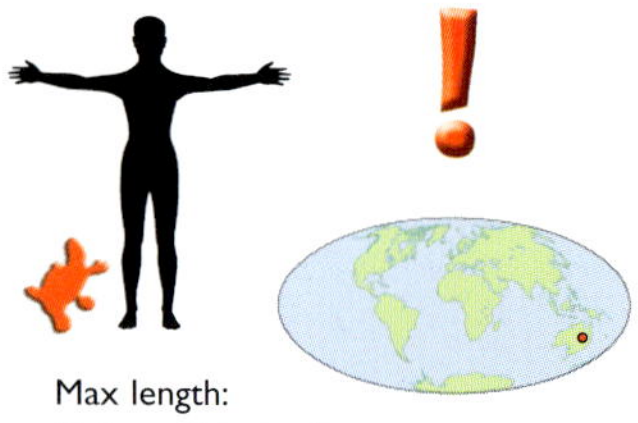

Max length:
56 cm (1 foot 10 inches)

Platypus

The platypus is the world's strangest animal. It looks so unusual that when it was discovered, some scientists thought it was a hoax. The platypus lays eggs and is a mammal. It lives along the banks of streams and rivers, using its duck-like bill to search for food along the bottom. This rare animal is threatened by habitat destruction and water pollution.

Fact

The platypus is one of only a few mammals that use poison. Males have a poisonous spur on their hind feet.

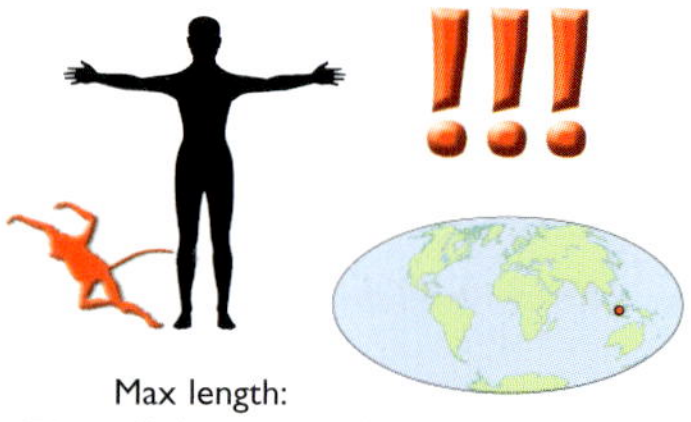

Max length:
76 cm (2 feet 6 inches)

Proboscis monkey

The proboscis monkey is easily identified by its long nose (proboscis). It is found only in mangrove forests near fresh water on the island of Borneo. The proboscis monkey is endangered by the destruction of its forest home.

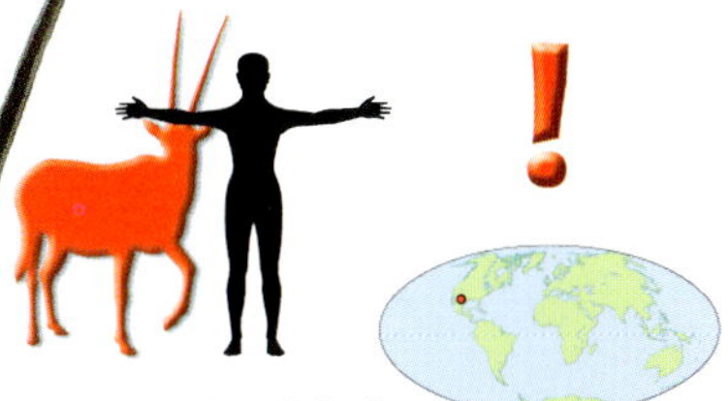

Max length: 1.5 m (5 feet)

Pronghorn

The pronghorn is a unique North American mammal. The peninsula pronghorn subspecies, found only in Baja California, has nearly died out. The main reasons for its decline are habitat loss and hunting. It is estimated that there are only 150 peninsular pronghorn, 475 sonoran pronghorn and less than 300 American pronghorn left in the wild.

Fact
The pronghorn is one of the fastest animals over distance, and can average about 65 km/h (40 mph) for several minutes.

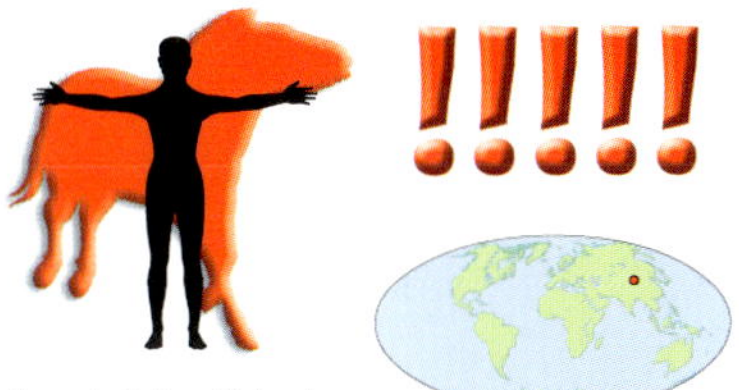

Max length: 2.1m (7 feet)

Przewalski's horse

Przewalski's horse is much smaller than domestic horses at 122-142 cm (48-56 inches). This horse has not been seen in the wild since 1968. About 1,200 Przewalski's horses can be found in zoos, private preserves, and protected areas in Mongolia. Small groups of captive-bred horses have been released into the wild.

Qq

Max length: 1.8 m (6 feet)

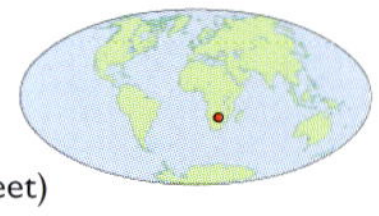

Quagga

The quagga was once believed to be a separate species that was hunted to extinction in the 19th century. Recent DNA studies show that the quagga is really a brown-and-white type of plains zebra. Some zoos and nature reserves in South Africa have produced a similar animal, and a quagga-like creature will soon be reintroduced into the wild.

Max wingspan: 1 foot 33 cm (1 inch)

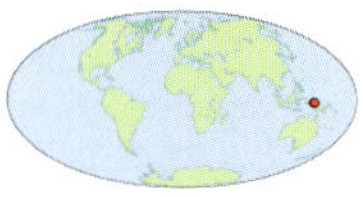

Queen Alexandra's birdwing butterfly

With its massive wingspan, the Queen Alexandra's birdwing butterfly is the largest butterfly in the world. This magnificent insect is highly prized by collectors, but strictly protected by law. It is endangered because of the destruction of the rainforest. Captive-breeding programmes have allowed some butterflies to be released back into the wild.

Max length: 40 cm (1 foot 4 inches)

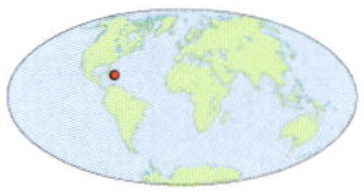

Quetzal

The resplendent quetzal lives high in the canopy of Central America's rainforests. This beautiful bird is hunted for its long, brightly coloured feathers. The ancient Aztec and Mayan civilizations respected them and made quetzal hunting punishable by death. Habitat destruction continues to threaten this gorgeous species.

Max legspan:
12.7 cm (5 inches)

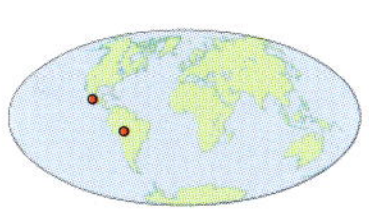

Red-kneed tarantula

The red-kneed tarantula is a large, colourful spider that has become very popular as a pet. Tarantulas are slow to breed in captivity. Hundreds of thousands have been taken from the wild, where they are becoming increasingly rare. The export of red-kneed tarantulas is now banned by law but many are still smuggled out illegally for sale in pet shops.

Fact
The largest tarantulas have a legspan of almost 30 cm (12 inches).

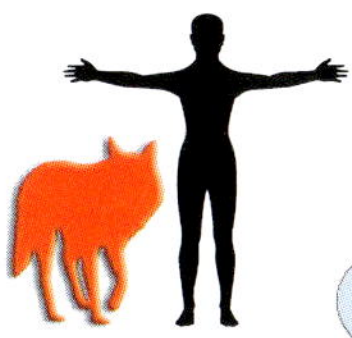

Max length:
1.3 m (4 feet 3 inches)

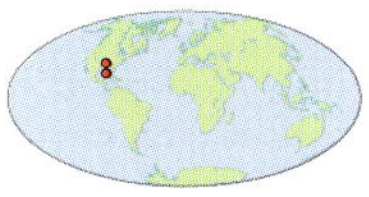

Red wolf

The red wolf became extinct as a wild animal during the 1960s. This was the result of centuries of persecution by farmers who believed that packs of wolves attacked their cattle, horses and sheep. Since 1987, groups of red wolves raised in captivity have been released throughout the eastern USA. So far, these animals appear to be successfully readapting to life in the wild.

Rr

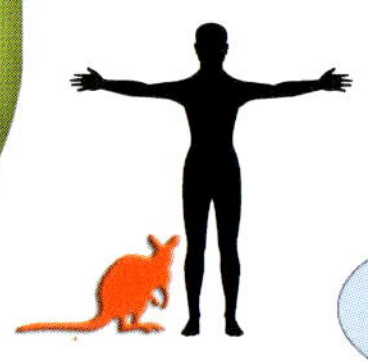
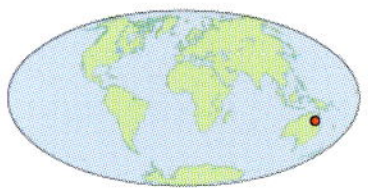

!!!

Max length:
64 cm (2 feet 1 inch)

Rock wallaby, Prosperine

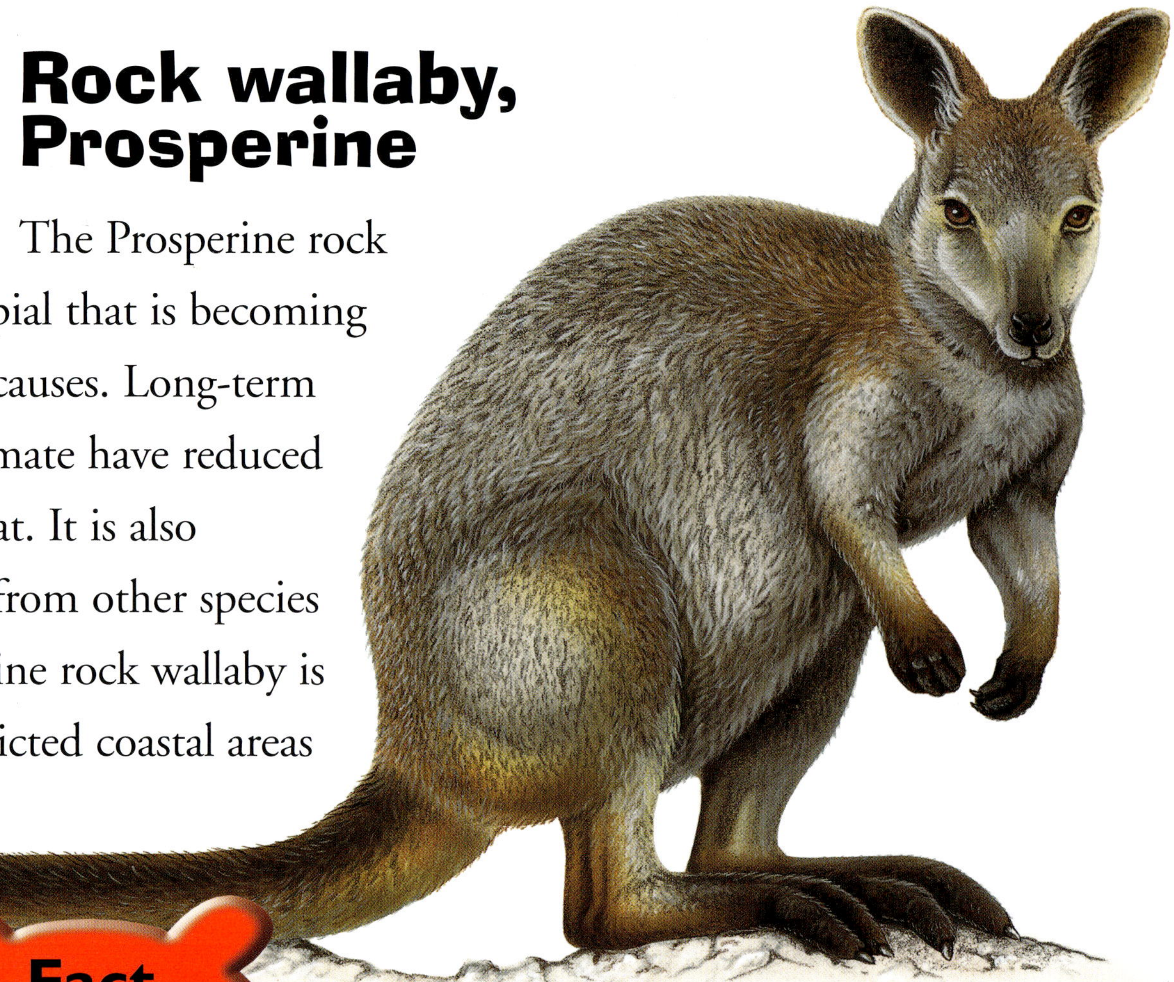

The Prosperine rock wallaby is a dog-sized marsupial that is becoming extinct as a result of natural causes. Long-term changes to the Australian climate have reduced its grassland and forest habitat. It is also facing competition for food from other species of rock wallaby. The Prosperine rock wallaby is now found only in very restricted coastal areas and on a few small islands.

Fact
The Prosperine rock wallaby is so rare that it was not scientifically documented until the 1970s.

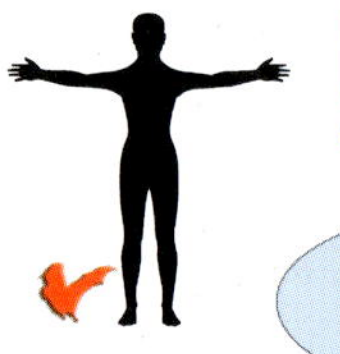
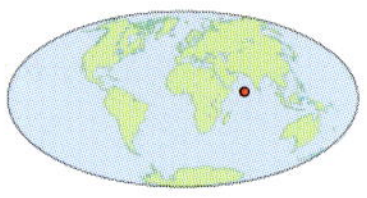

!!!!

Max length:
36 cm (1 foot 2 inches)

Rodriguez fruit bat

The Rodriguez fruit bat – also known as a flying fox – is found on the small island of Rodriguez in the middle of the Indian Ocean. Its numbers have been reduced by logging operations that have destroyed its forest habitat, poisoning, and electrocution from power lines. Captive-breeding programmes have been established to try to save this unique island species.

Max length:
61 cm (2 feet)

Ruffed lemur

The ruffed lemur, like all other lemurs, is found only in Madagascar. It is the largest lemur at 3.5-4.5 kg (7-10 pounds). The ruffed lemur is a tree-dweller that jumps from tree to tree. It faces many dangers: hunting, trapping, and the destruction of the tropical rainforests for timber.

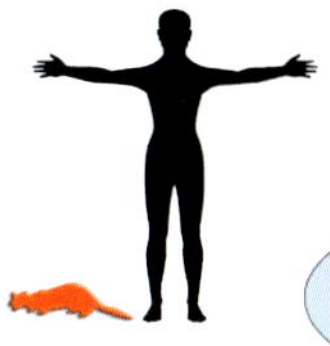

Max length:
46 cm (1 foot 6 inches)

Sable

The sable is a member of the weasel family that has very thick, warm fur. Throughout history, the sable has been hunted nearly to extinction for its exquisite fur. It is now protected by law. Most of the sable skins used today come from animals raised in captivity. Illegal hunting still threatens the remaining wild populations.

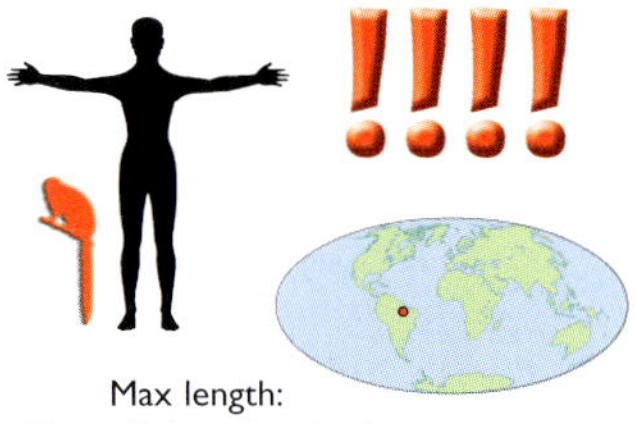

Saki

Max length:
48 cm (1 foot 7 inches)

The monk saki is one of seven species of saki monkey, all of which are endangered by loss of habitat. This very shy monkey feeds on seeds, nuts and insects. The monk saki lives in small family groups and spends most of its time high in the rainforest canopy. Sakis do not survive well in captivity.

Fact
The monk saki is also known as the red-bearded saki because of the reddish hair beneath its chin.

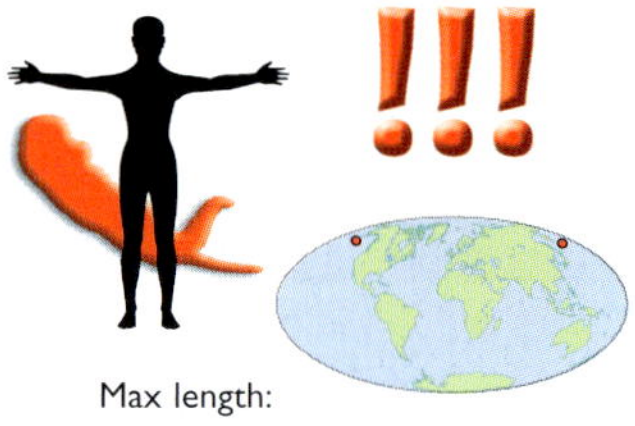

Sea otter

Max length:
1.3 m (4 feet 2 inches)

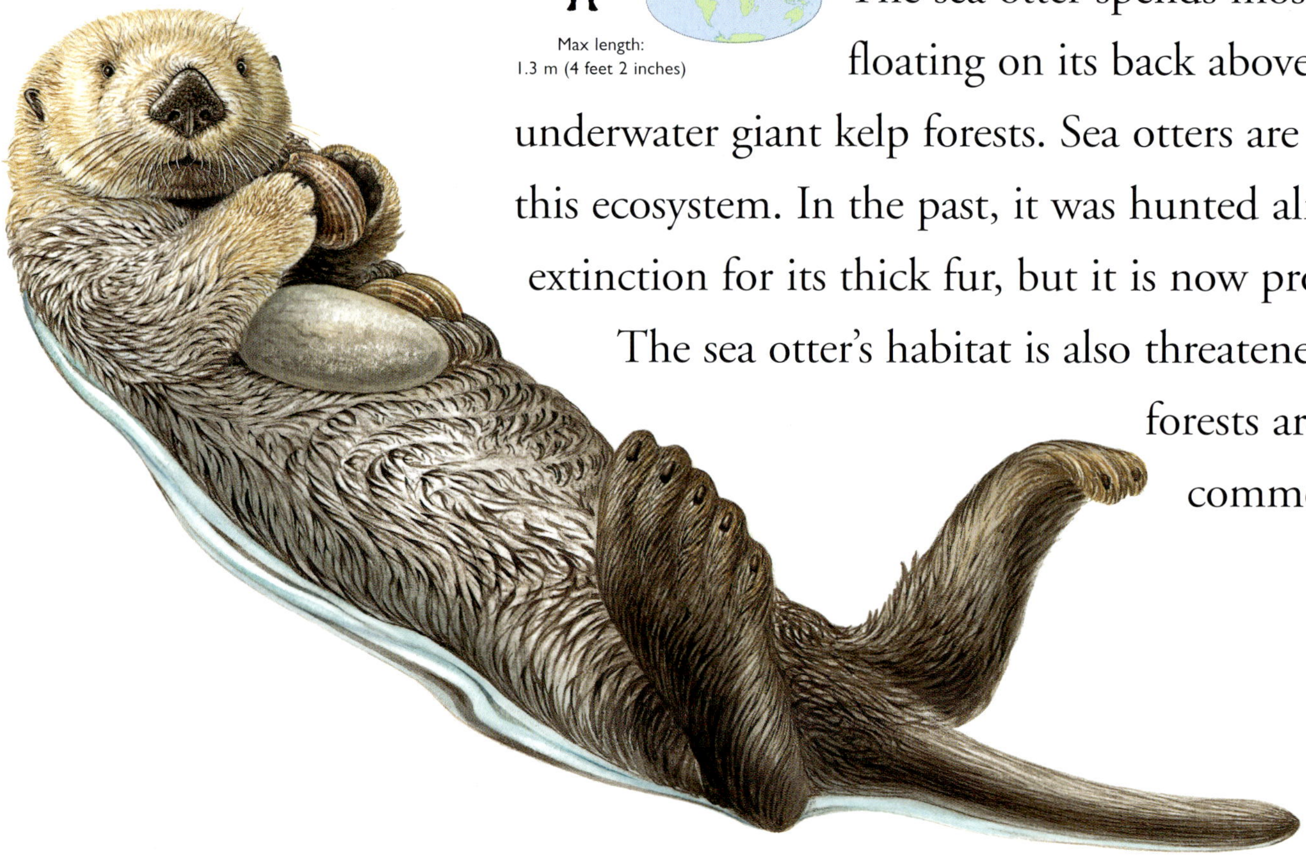

The sea otter spends most of its time floating on its back above the underwater giant kelp forests. Sea otters are key species for this ecosystem. In the past, it was hunted almost to extinction for its thick fur, but it is now protected by law. The sea otter's habitat is also threatened as the kelp forests are harvested for commercial purposes.

Max length:
51 cm (1 foot 8 inches)

Sloth

The Brazilian three-toed sloth is related to anteaters and armadillos. It spends most of its time hanging upside down from tree branches. The Brazilian three-toed sloth moves very slowly and is often killed during logging operations. Efforts are being made to collect sloths before logging, but this gentle animal does not do well in captivity.

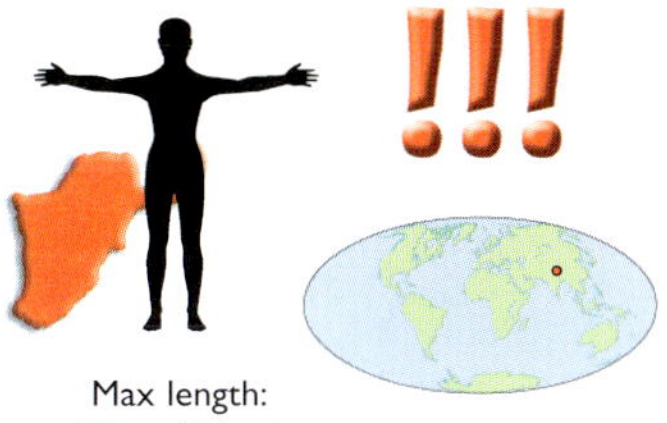

Max length:
1.5 m (5 feet)

Snow leopard

The snow leopard lives in the cold, remote alpine meadows and forests of the Himalayan plateau. Its thick, woolly fur is highly prized by hunters. Illegal hunting continues to threaten this species. Many captive-breeding programmes are trying to save this cat from extinction as numbers continue to diminish.

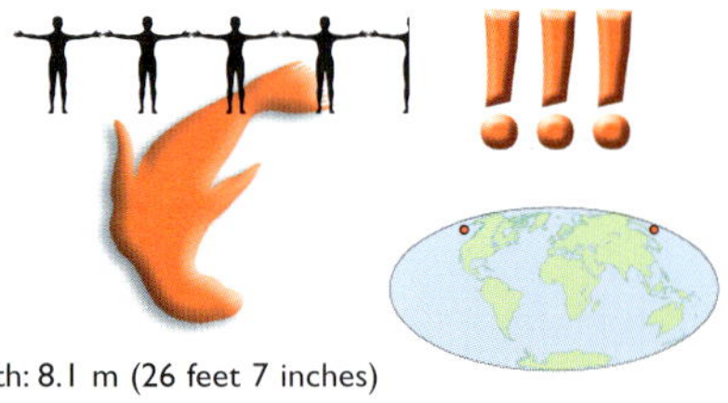

Max length: 8.1 m (26 feet 7 inches)

Steller's sea lion

Steller's sea lion is the largest sea lion and an adult male can weigh more than 1,000 kg (2,200 pounds). It is a ferocious predator, hunting underwater for fish, seals and otters. It is endangered by increased human activity around the shores of the northern Pacific. Its pups are hunted for their glossy black fur.

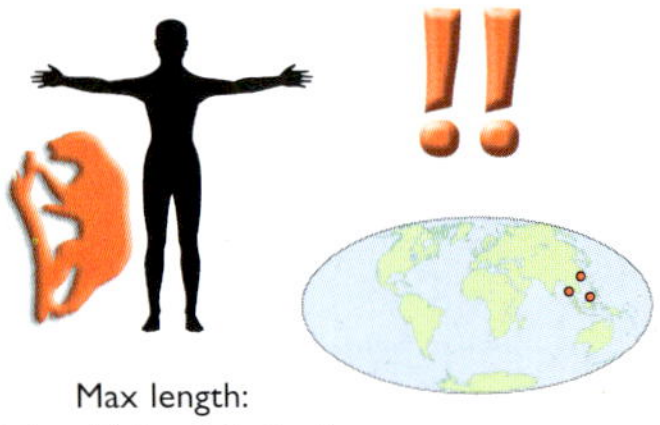

Max length: 1.4 m (4 feet 6 inches)

Sun bear

The sun bear is the smallest and rarest bear, and the only one to live in tropical forests. Long, curved claws help it climb trees in search of eggs, fruit and wild honey, giving it the name 'honey bear'. The sun bear's continued existence is under threat because logging operations are destroying its forest home.

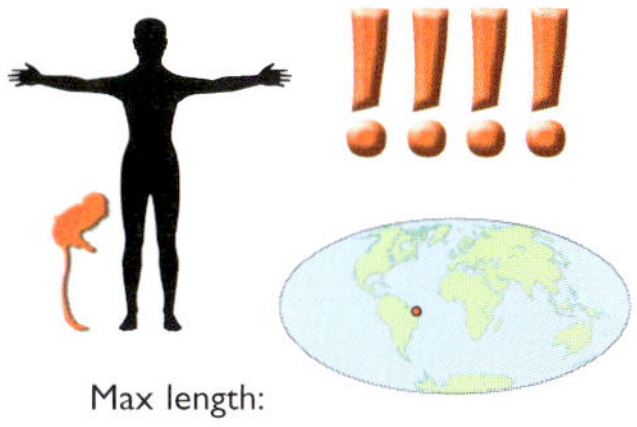

!!!!

Tamarin

Max length:
33 cm (1 foot 1 inch)

The golden lion tamarin has mane-like fur and is found only in one small area of Brazil's coastal forest. In the middle of the 20th century, this unusual-looking monkey almost became extinct when its habitat was cleared for agriculture. There are only 400 left, but captive-breeding programmes may save this species.

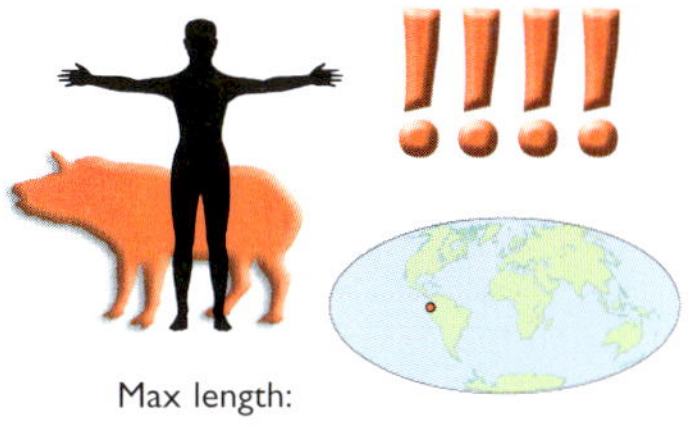

!!!!

Tapir

Max length:
1.8 m (6 feet)

The mountain tapir lives only in the high forests of the northern Andes. It is a reclusive animal that spends most of the day hiding in thickets and shadows. The mountain tapir is critically endangered by loss of habitat. It is estimated that they will become extinct in the next 20 years.

Tt

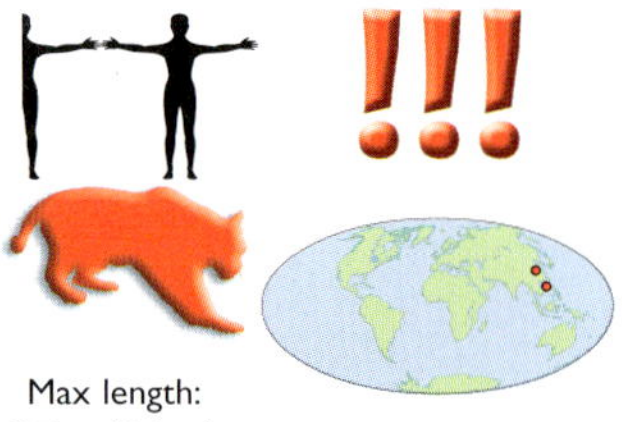

Max length:
2.7 m (9 feet)

Tiger

The tiger is the world's largest cat and one of the most endangered. There are five surviving subspecies (50 years ago there were eight!), of which the Siberian tiger is the rarest, with fewer than 200 left in the wild. Although they are protected by law, tigers are still threatened by habitat destruction and poaching – both for their skins and bones, which are used in traditional Asian medicines.

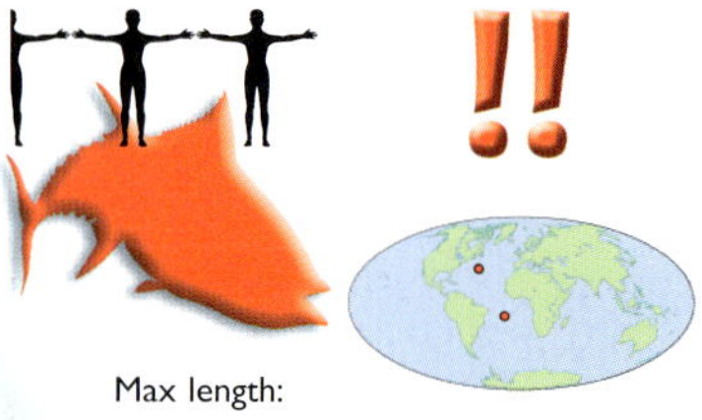

Max length:
4.6 m (15 feet)

Tuna

The northern bluefin tuna has become the world's favourite fish to eat – so much so that it is now in danger. Although these fish are still found in large schools, their numbers are decreasing rapidly. Without conservation programmes, this tuna will be extinct within five years.

Max length:
58 cm (1 foot 11 inches)

Uakari

The bald uakari has a distinctive red face and is found only in isolated, swampy regions of the Amazonian rainforest. It has long been hunted by local people as a source of food and bait. Its remote habitat is threatened with destruction. A combination of logging operations, agriculture and river-control projects are draining and clearing the forest swamps where the uakari lives.

Fact
The vaquita can survive in lagoons so shallow that its back can be seen clearly above the water's surface.

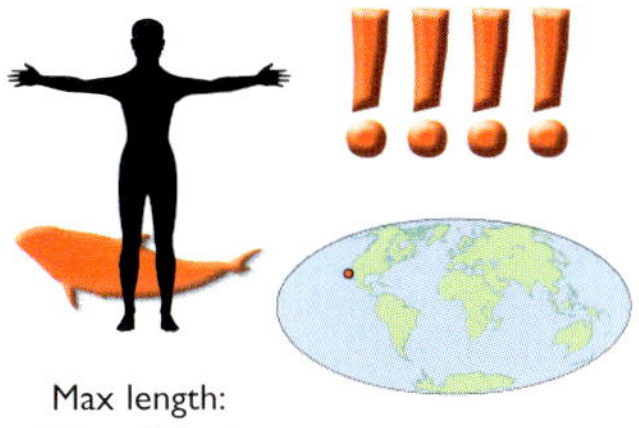

Max length:
1.5 m (5 feet)

Vaquita

The vaquita, also known as the Gulf of California porpoise, is on the brink of extinction. This rare marine mammal is often mistaken for the totoabo fish, which is a delicacy. It is threatened by water pollution, fishing nets and collision with boats. There are now laws in place to save the vaquita, but it may be too late to protect it in the wild.

Vv

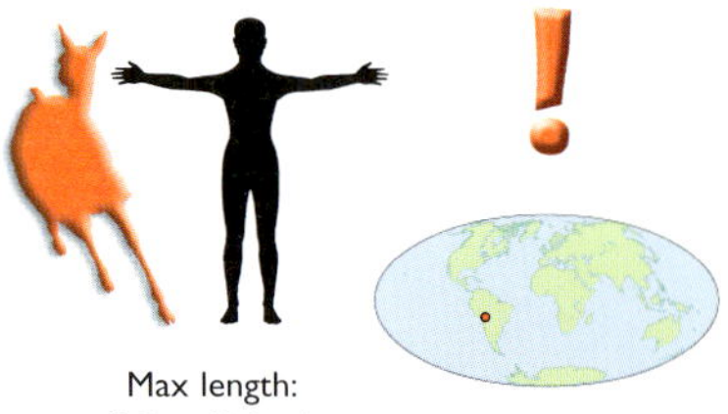

Max length:
1.8 m (6 feet)

Vicuña

The vicuña is the ancestor of the domesticated alpaca, and is related to the camels of Asia and Africa. Its coat is made of very fine, warm fibres, for which the vicuña had been hunted almost into extinction. By the 1960s, it became protected by law. Thanks to conservation measures, vicuña numbers are up to about 125,000, with most of them now living in specially established national parks.

Fact
Vicuñas live on dry, windswept grassland more than 3,600 m (12,000 feet) above sea level.

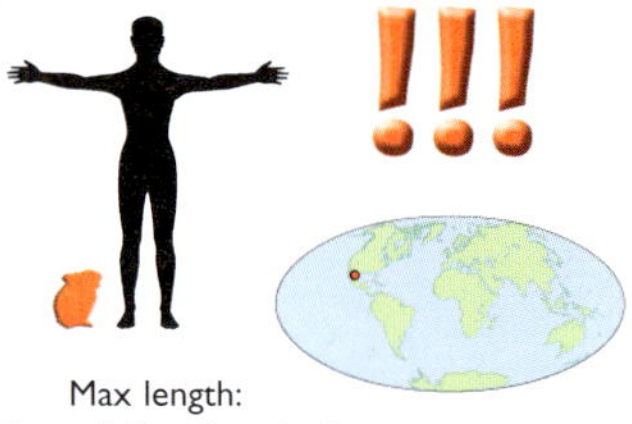

Max length:
36 cm (1 foot 2 inches)

Volcano rabbit

The volcano rabbit is found only in the pine forests near the peaks of several inactive volcanoes near Mexico City. Here, its dark fur blends well with the volcanic soil. This very restricted habitat is now under pressure from logging and over-grazing by cattle and sheep.

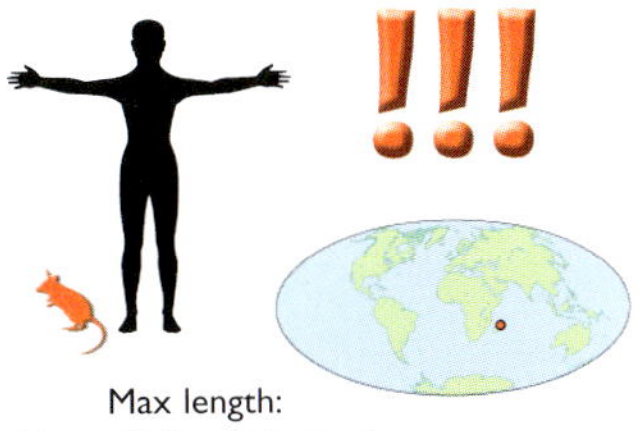

Max length:
36 cm (1 foot 2 inches)

Votsota

The votsota, which is also known as the Malagasy giant rat, is found only on the island of Madagascar. It grows to be more than 60 cm (24 inches) from nose to tail. The plant-eating votsota is threatened by climate changes and destruction of its forest habitat.

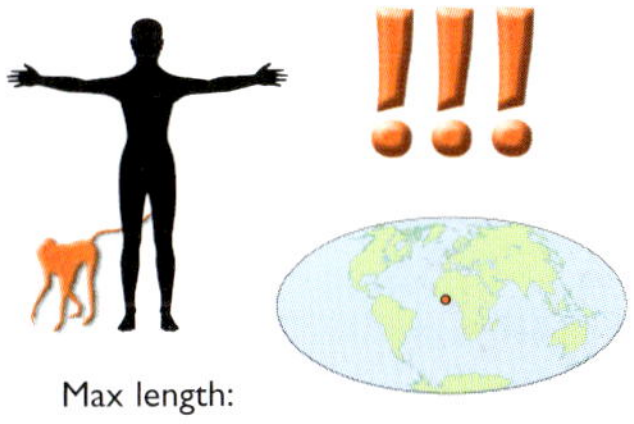

Max length:
61 cm (2 feet)

White-collared mangabey

The white-collared mangabey is now found only in a few restricted areas of West African coastal forest. Its numbers have been drastically reduced by hunting, forest clearing, and capture for the pet trade and medical experiments.

Max height:
1.4 m (4 feet 5 inches)

Whooping crane

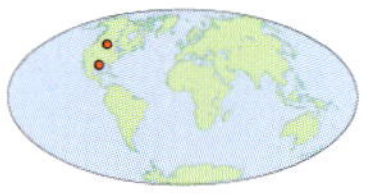

The whooping crane is now at considerable risk, with less than 100 remaining in the wild. In spite of strict protection, its numbers continue to decline. This bird migrates from Canada to spend each winter in small areas of Texas and Florida. However, coastal development has destroyed most of its natural wintering grounds. The whooping crane now survives only in nature reserves.

Fact
The whooping crane stands 1.5 m (5 feet) tall and has a wingspan of about 2.1 m (7 feet).

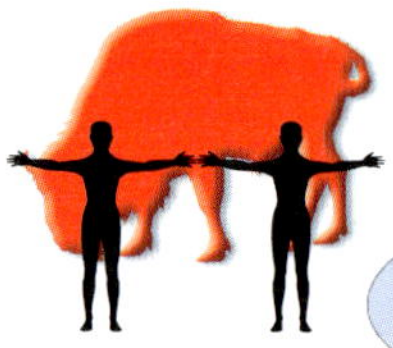

Max length:
3.4 m (11 feet)

Wisent

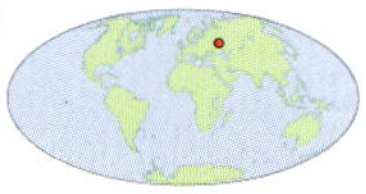

The wisent, or European bison, became extinct in the wild about 100 years ago. It is thought to be closely related to the buffalo (American bison). The wisent survived in captive herds and has been released into Bialoweiza Forest nature reserve in Poland and Belarus. Its main threat is now from illegal hunting.

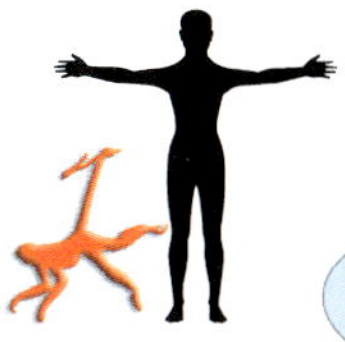

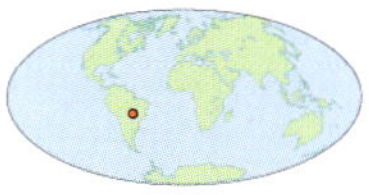

Max length:
58 cm (2 feet)

!!!

Woolly spider monkey

The woolly spider monkey, which is also known as the muriqui, has become endangered as a result of the clearance of Brazil's coastal forests. The woolly spider monkey only eats the leaves of certain trees, so when its natural habitat was destroyed, its only source of food disappeared. Fewer than 500 of these monkeys remain in the wild, protected in nature reserves.

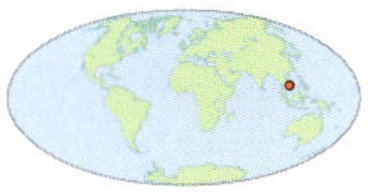

Max length:
8 cm (3 inches)

!!!!

Xenopoecilus

This small freshwater fish with a big name is found only in Lake Poso in Indonesia and is on the brink of extinction. More simply known as the Sarasin's minnow, it is threatened by the introduction of non-native fish and poor water quality. Captive-breeding programmes are showing some success.

Max length:
3.4 m (11 feet)

!!

Yak

The wild yak, which is larger than its domesticated cousin, lives on the bleak, snow-swept mountains of the Himalayas. It is estimated that there are around 15,000 of these animals remaining in the wild. Herds of wild yak are a tempting target for hunters, and law enforcement is difficult in the cold and remote habitat.

Max length:
71 cm (2 feet 4 inches)

!!

Yellow-throated marten

The yellow-throated marten was once common in eastern Asia. Habitat destruction and persecution by farmers (who see it as a pest because it preys on small deer) have greatly reduced its numbers. Taiwan's yellow-throated marten has joined the endangered list.

Max wingspan:
30 cm (1 foot)

!!!

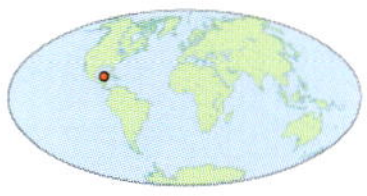

Zapata wren

The Zapata wren is a small bird that is found only at one end of the Zapata Swamp in Cuba. This bird seems to hate flying, so it makes an easy meal for predators. After brush fires burned large areas of its natural habitat, it was put on the critically endangered list in 1974. Happily, in 1998 this status changed after many more Zapata wrens were found.

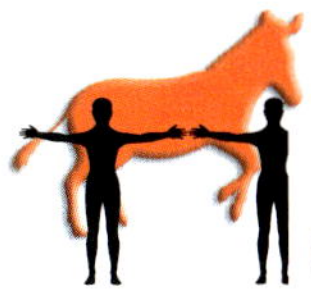

Max length:
3 m (9 feet 8 inches)

!!!

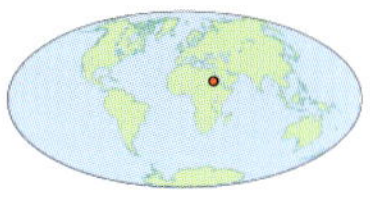

Zebra, Grevy's

Grevy's zebra, largest of the three zebra species, is found only in one small area of the East African Rift Valley. There was a high demand for zebra meat and skins, drastically reducing the population. There are only about 5,000 remaining. Hunting restrictions have allowed zebra numbers to recover a little, but pasture is scarce for the remaining population.

Glossary

Amphibian An air-breathing animal with a backbone that lays its eggs in water. Frogs, toads, newts and salamanders are the most commonly encountered types of amphibian.

Golden frog

Antelope One of a group of hoofed mammals. Antelope have slim legs and unbranched horns that are a permanent growth (unlike the horns and antlers of deer that are shed and regrown every year). The American pronghorn is often called an antelope.

Aquatic Describes an animal that spends most, or all, of its time in water.

Asteroid A lump of rock, much smaller than a planet, that orbits the sun. There are many thousands of asteroids, and Earth is struck by one about every 50–100 million years.

Captive breeding Breeding wild animals in captivity (in zoos or wildlife parks) with the intention of releasing them back into the wild.

CITES Convention on Trade in Endangered Species, a conservation treaty that prohibits the international trade and transportation of any animals (or parts of animals) that are endangered.

Cloud forest Region of damp forest found on the upper slopes of mountains that are permanently covered by cloud.

Conservation Preserving wild animals in their natural environment. Conservation measures include physical protection, such as fences around wildlife refuges; environmental protection, such as preventing pollution; and research into wild animal behaviour.

DDT A chemical that was widely used to kill insects in the 20th century. DDT was supposed to be harmless to other wildlife, but in fact was deadly, especially to birds. The use of DDT is now banned in most countries.

Dinosaur One of a group of reptiles that were the dominant land animals until they became extinct about 65 million years ago.

DNA A chemical found in every cell of every living thing. Each species has its own version of DNA that contains a coded blueprint for making that particular animal or plant.

Orangutan

Glossary

Dodo A flightless bird that lived only on the Indian Ocean island of Mauritius, where it had no natural enemies. Within about 20 years of being discovered by European sailors in the 17th century, the dodo was hunted to extinction.

Domesticated Describes a species that was once wild, but has been changed into a different species through many hundreds of years of being used as a farm animal or pet.

El Niño A climatic event that happens about every 7 to 11 years in the Pacific Ocean region. During an El Niño year, the normal weather patterns are disrupted. Storms and torrential rainfall can cause flooding and mudslides in some places, while other places suffer drought and brush fires.

Endangered Describes a species that has such a small remaining population that it is in danger of becoming extinct.

Estuary The lowest part of a river where it enters the sea. The water in an estuary is a mixture of fresh and salty water. Some species can only live in these special conditions.

Extinct Describes a species that no longer exists.

Fresh water Rainwater, river water and the water in most lakes is called fresh water because it contains no salt.

Quagga

Fungi A form of life that is neither plant nor animal. Fungi grow underground and on decaying vegetation. Mushrooms and mould are the most commonly encountered types of fungi.

Habitat The combination of landscape, climate, vegetation and animal life that form the natural environment for a particular species.

Black rhinoceros

Horn A hard growth (sometimes with a bony core) that extends from the head of some mammals. Also, the substance these growths are made of, which is similar to human hair or fingernails.

Insect A small invertebrate animal with six legs; many species have wings. Insects, along with crabs, lice, shrimp, spiders and scorpions, belong to the arthropod group. There are more insect species on Earth than all other animal and plant species put together.

Introduced species A type of 'living pollution' – animals (and plants) that do not occur naturally in a particular habitat, but are there because of human activity. Introduced species include escaped farm animals (for example, pigs and goats), pets (such as cats and dogs) and vermin (rats and mice).

Glossary

Invertebrate An animal that does not have an internal skeleton with a backbone. Insects, slugs, snails, spiders and worms are all invertebrates.

Ivory Illegal animal product obtained from the tusks of elephants, walruses and narwhals, that was once widely used for jewellery, musical instruments and decorative items.

Kelp A seaweed that grows in long strands anchored to the sea bottom. Along some coasts, kelp forms 'forests' that attract many sea creatures.

Krill Small shrimp-like animal that lives in vast schools in the Southern Ocean around Antarctica. It is the main food source for many whales.

Living fossil A species that has survived for many millions of years after similar species became extinct.

Mammal A warm-blooded, vertebrate animal that produces live-born young. Most mammals are covered with hair and live on land. There are a few aquatic mammals, such as seals, dolphins and whales.

Coelacanth

Mangrove A tropical tree that can survive with its roots in shallow water. Mangrove forests grow along many tropical coastlines and provide a sheltered habitat for many animals.

Marsupial One of a group of mammals that raise their young in a pouch at the front of the female's body. Marsupials are found only in Australia and the Americas.

Migration Regular movement of animals from one habitat to another. Some birds migrate thousands of miles each year.

Pest Any animal (or plant) that is harmful to human activity. Farmers consider many wild animals to be pests because they eat crops or attack livestock.

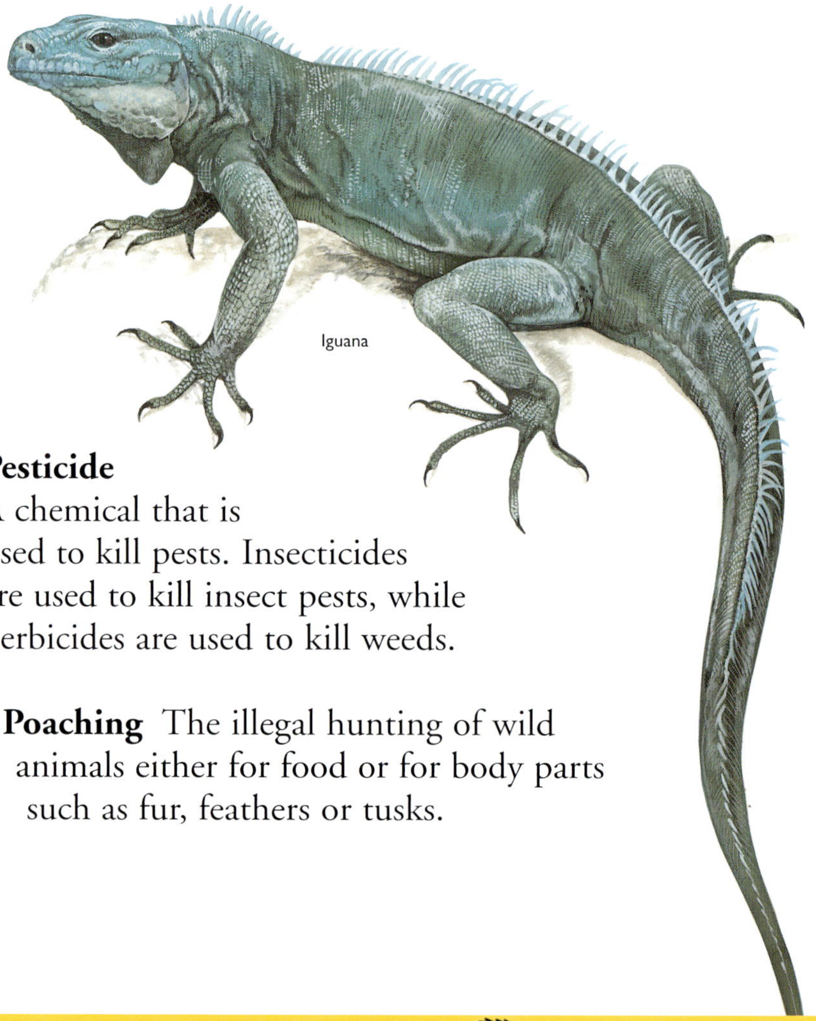

Iguana

Pesticide
A chemical that is used to kill pests. Insecticides are used to kill insect pests, while herbicides are used to kill weeds.

Poaching The illegal hunting of wild animals either for food or for body parts such as fur, feathers or tusks.

Glossary

Pollution The presence of unnatural substances in a habitat. Pollution is the result of human activity. Air pollution is caused by fumes from vehicle exhausts and smoke from factories. Water pollution is mainly caused by industrial and household waste, and also by chemicals used in farming.

Predator An animal that hunts and eats other animals.

Prehensile Describes a tail that an animal can control like a fifth limb. Some tree-dwelling animals have prehensile tails, which they use for holding on to branches when climbing in the trees.

Primate One of a group of mainly tree-dwelling mammals that have the same basic body pattern as human beings. According to scientific classification, humans are members of the primate group.

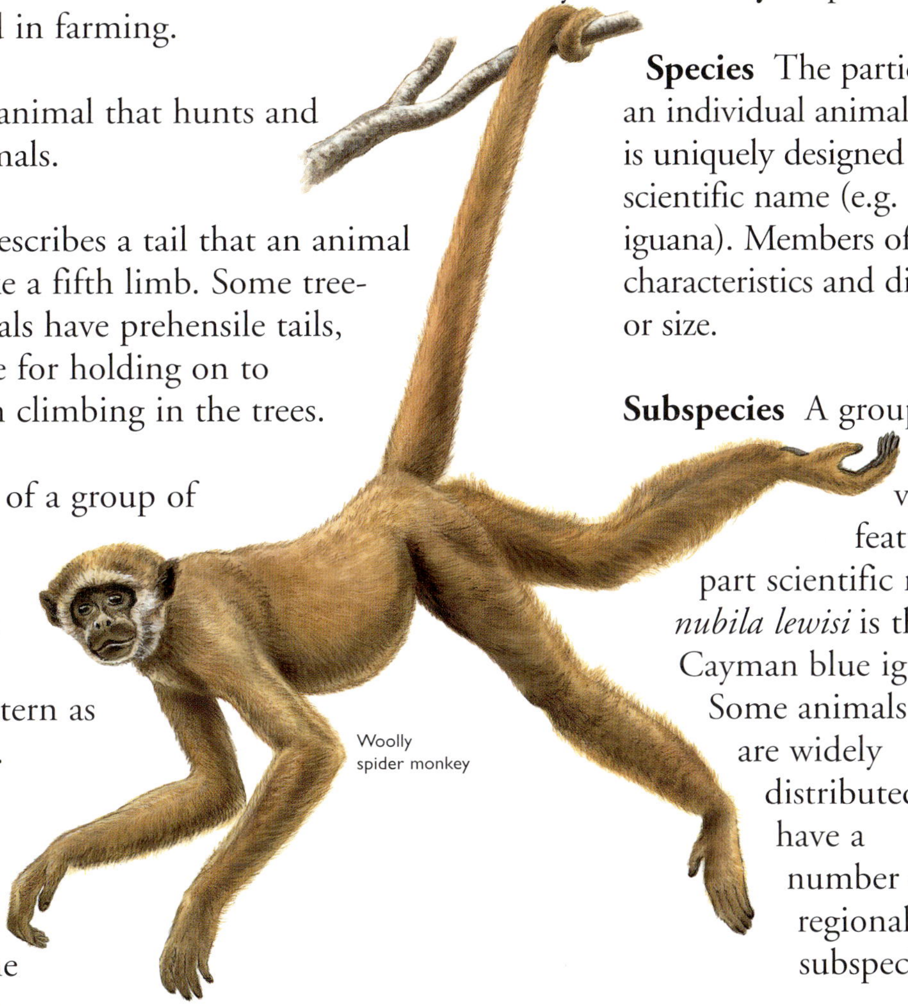

Woolly spider monkey

Prey An animal that is hunted and eaten by others.

Refuge A place of safety. Wildlife refuges range in size from vast national parks to small 'green' areas in towns and cities.

Reptile A cold-blooded vertebrate animal that breathes air and produces eggs. Some do produce live young. Crocodiles, lizards, turtles, tortoises and snakes are all reptiles.

Rodent One of a group of small mammals that includes mice, rats and squirrels, but not shrews or rabbits.

Sanctuary A place of safety.

Species The particular scientific group to which an individual animal (or plant) belongs. Each species is uniquely designed for life and has a two-part scientific name (e.g. *Cyclura nubila* is the ground iguana). Members of the same species all share the same characteristics and differ only slightly in colouration or size.

Subspecies A group of animals belonging to the same species that share the same variation in their characteristic features. Each subspecies has a three-part scientific name (e.g. *Cyclura nubila lewisi* is the grand Cayman blue iguana). Some animals that are widely distributed have a number of regional subspecies.

Asian elephant

Tropical Belonging to the geographical region around the equator, between the Tropic of Cancer and the Tropic of Capricorn. The tropical climate is usually hot and rainy.

Tusk An extended tooth that projects from the head of some mammals, such as elephants and walruses. Animal tusks are the source of ivory, which some people consider valuable.

Index